Geographical Diversions

EST. 75 1938 YEARS
THE UNIVERSITY OF GEORGIA PRESS 2013

GEOGRAPHIES OF JUSTICE AND SOCIAL TRANSFORMATION

Geographical Diversions

TIBETAN TRADE, GLOBAL TRANSACTIONS

TINA HARRIS

THE UNIVERSITY OF GEORGIA PRESS
Athens and London

© 2013 by the University of Georgia Press

Athens, Georgia 30602

www.ugapress.org

All rights reserved

Designed by Kaelin Chappelle Broaddus

Set in Minion Pro and Trade Gothic by Graphic Composition, Inc., Bogart, Georgia

Printed and bound by Thomson-Shore

The paper in this book meets the guidelines for permanence and durability of the Committee on Production Guidelines for Book Longevity of the Council on Library Resources.

Printed in the United States of America

17 16 15 14 13   P   5 4 3 2 1

Library of Congress Cataloging-in-Publication Data

Harris, Tina, 1976–

    Geographical diversions : Tibetan trade, global transactions / Tina Harris.

        p. cm. — (Geographies of justice and social transformation)

    Includes bibliographical references and index.

    ISBN 978-0-8203-3866-8 (hardcover : alk. paper) — ISBN 0-8203-3866-4 (hardcover : alk. paper) — ISBN 978-0-8203-4512-3 (pbk. : alk. paper) — ISBN 0-8203-4512-1 (pbk. : alk. paper)

    1. Ethnology—China—Tibet Autonomous Region. 2. Commerce—China—Tibet Autonomous Region. 3. Sheepherding—Economic aspects—China—Tibet Autonomous Region. 4. Human geography—China—Tibet Autonomous Region. 5. Tibet Autonomous Region (China)—Boundaries. 6. Tibet Autonomous Region (China)—Economic conditions. I. Title.

GN635.C5H37 2013

305.800951'5—dc23

2012042286

British Library Cataloging-in-Publication Data available

To my grandfather, Isao Honjo, who told me that
when I grew up, I would write a book.

CONTENTS

FIGURES

ACKNOWLEDGMENTS

It seems unfair that single-authored books like this are considered by default the product of one individual. In this case, nothing could be farther from the truth. I doubt I would have been able to complete this project without the support, cooperation, camaraderie, and expertise of numerous friends and colleagues. But before I begin to name specific names, the most gratitude goes out to the many traders, shopkeepers, and friends in Tibet, India, and Nepal who answered my questions, many of whom I cannot name for sensitive reasons. Your generosity was unprecedented: thank you.

The seedlings of this project began to take root on my first trip to Tibet in 1997, but the bulk of the research was conducted between 2005 and 2007. I am grateful to various institutions, including the Tibetan and Himalayan Library at the University of Virginia, the Tibet Academy of Social Sciences, the Centre for Nepal and Asia Studies at Tribhuvan University, and the International Trust for Traditional Medicine for facilitating portions of my research on material culture and old trading practices.

I am extremely grateful for the financial support I have received during the research and writing period, which was made possible by the Wenner-Gren Foundation for Anthropological Research, the CUNY Center for Place, Culture, and Politics, the Great Issues Forum, and the B. Altman Dissertation Fellowship. In addition, two summers at the British Library, provided by the Map Library's Helen Wallis Fellowship, were invaluable for historical research and for writing. I would especially like to thank Peter Barber and the Map Library staff, as well as Emma Jackson and Adey Lobb for their warm hospitality and proper tea.

Friends and colleagues affiliated in various ways with the world of Tibet and Himalayan studies have provided crucial advice, encouragement, and shared conversations over good (and occasionally not-so-good) meals. Some of the many people I wish to thank include Anna Balikci-Denjongpa, John Bray, Catherine Dalton, Hubert and Nazneen Decleer, Brandon Dotson, Kristina Dy-Liacco, David Germano, Clare Harris, Lauran Hartley, Kabir Heimsath, Sandip Jain, Carole McGranahan, Peter Moran, Tenzin Norbu, Annabella Pitkin, Andy Quintman, Prista Ratanapruck, Leigh Sangster, Tsering Shakya, Sara Shneiderman, Megan Sijapati, Norbu Tsering, Mark Turin, Pema Wangchuk, Eveline Yang, and Emily Yeh. Sincere gratitude goes out to Tsam-la, who provided indispensable help during the fieldwork component. I would like to extend extra-special thanks to Kimberly Dukes, Chris Hatchell, and especially Carl Yamamoto.

Along the way, several families in Lhasa, Kathmandu, and Kalimpong were overly generous with their time and conversations. I especially thank Jigme and his family for taking me in and sharing their stories; Tenzin and family, members of the Tuladhar family, Maimuna, Shabnam, and family, and Razia and Aziz Baba for their friendship. Please note that the views put forth in this book do not necessarily overlap with the views of those mentioned in these acknowledgments, and that I alone am responsible for any errors.

I owe much gratitude to friends and colleagues in New York and beyond who have offered comments and advice on various incarnations of this project, especially Itty Abraham, Gregory Donovan, Jen Gieseking, Bina Gogineni, Elizabeth Johnson, Banu Karaca, Duncan McDuie-Ra, John Morrissey, Nandini Sikand, Malini Sur, and Anand Vivek Taneja. In particular, conversations at the CUNY Graduate Center Anthropology Department, the CUNY Center for Place, Culture, and Politics Seminar on Geopolitics and Insecurity, and the Antipode Summer Institute were a source of inspiration, intellectual camaraderie, and joy. I also thank the anthropology department at Texas Tech University for their generosity during my brief year and a half there.

Sections of this book have been published elsewhere: much of chapter 2 appears as "From Loom to Machine: Tibetan Aprons and the Configuration of Space" in *Environment and Planning D: Society and Space* 30(5): 877–95, 2012 (Pion Ltd., London, www.pion.co.uk); parts of chapter 3 have been published as "Silk Roads and Wool Routes: Contemporary Geographies of Trade between Lhasa and Kalimpong" in *India Review* 7(3): 200-222, 2008 (reprinted by permission of Taylor & Francis Ltd., http://www.tandf.co.uk/journals); and chapters 4 and 5 contain sections that have been published in *Political Geography*, Tina Harris, "Trading Places: New Economic Geographies across Himalayan Borderlands," 2013, http://dx.doi.org/10.1016/j.polgeo.2012.12.002, with permission from Elsevier Science Ltd.

Thanks to Robbie Barnett, Michael Blim, David Harvey, and Jane Schneider, who patiently guided me through the various stages of my PhD dissertation. I couldn't have asked for a better PhD supervisor, mentor, and dear friend than the peerless Neil Smith. Neil challenged my thinking on all levels while believing in my work every step of the way (even when I didn't). The devastating news of his death arrived just before the copyedits for this book were finalized, so it is particularly heartbreaking to have to thank him only in writing and not in person. Thank you so much, Neil. Your students will try to do you proud.

In Amsterdam, my new and delightful home, I thank the Moving Matters Research Cluster of the University of Amsterdam and Amsterdam Institute for Social Science Research, particularly Mario Rutten and Willem van Schendel, for comments on chapters and for general encouragement during my initial foray into academic life in the Netherlands.

Many thanks to the valiant editors, staff, and board members of the University of Georgia Press, especially Regan Huff, John Joerschke, Derek Krissoff, and Nik Heynen; to Doug Williamson at Hunter College for creating the maps; to Jennifer Reichlin for her copyediting expertise; to J. Naomi Linzer for the index; and to the two anonymous reviewers for their very insightful comments and suggestions.

My parents and family were my main source of encouragement, inspiration, and support from the very beginning. I simply couldn't have arrived here without them. Finally—thank you for being there for me, George.

A NOTE ON TRANSLITERATION

Tibetan, English, Nepali, and Chinese (Mandarin) were the main languages used in this project. Newari (also called Newah or Nepalbhasa), Hindi, and Marwari terms were also spoken by several individuals but are not quoted in this text. For Chinese words, I have used the standardized pinyin system for terms such as *hukou* (household registration system), but I use English phoneticizations without italics for more commonly used words such as Beijing. Similarly, for Tibetan, I have used the Wylie transliteration model for words such as *kha btags* (ceremonial scarves) but English phoneticizations without italics for words such as the Barkor (the central marketplace and pilgrimage circuit in Lhasa) or yak. Words in Tibetan are italicized on their own; words in Chinese or Nepali are italicized and marked by the abbreviations Cn. or Nep. (for example, Cn., *Xizang*; Nep., *thulo*). Interviews in Tibet were conducted in Tibetan, unless otherwise indicated. Interviews in India and Nepal were conducted in English, unless otherwise indicated.

Tibet, Trade, and Territory

You do not come to Euphemia only to buy and sell, but also because at night, by the fires all around the market, seated on sacks or barrels or stretched out on piles of carpets, at each word that one man says—such as "wolf," "sister," "hidden treasure," "battle," "scabies," "lovers"—the others tell, each one, his tale of wolves, sisters, treasures, scabies, lovers, battles. And you know that in the long journey ahead of you, when to keep awake against the camel's swaying or the junk's rocking, you start summoning up your memories one by one, your wolf will have become another wolf, your sister a different sister, your battle other battles, on your return from Euphemia, the city where memory is traded at every solstice and at every equinox.

Italo Calvino, *Invisible Cities*

Diversions

The temporal rhythms of cities are marked not just by seasonal changes but also by variations in trading practices. In the transition from autumn to winter in Lhasa, the tourists who buy trinkets from the marketplace vendors that are common in the height of summer begin to peter out until they are almost completely absent. The luxury hotels grow quieter. There is a momentary, almost palpable pause, and soon nomads from rural Amdo and Kham—other Tibetan-speaking regions on the plateau—begin to appear in the streets for trade and for pilgrimage, purchasing household items like blenders and blankets and sometimes bartering small trinkets for yak meat.

It was during one of these transitional moments, right before the city plunged into a deep winter—bitterly cold at night but clear and sunny during the day—that I began my fieldwork for this book. I was sitting indoors with Lobsang, an older man in his seventies who traded Tibetan wool and Indian foodstuffs between Lhasa and Kalimpong in the 1950s.[1] The house was dark and comfortable, and he seemed relaxed talking about the past, even in the tense political atmosphere of China's Tibet Autonomous Region (TAR). I felt like the interview was going well. Lobsang remembered quite a bit about his life, and I took notes on the kinds of items he traded, names for old systems of measurements, and the towns he passed through with his mule caravan. Toward the end of our discussion, I glanced at my list of interview questions and asked him if he could draw me a map of what he remembered of his journey from Lhasa to Kalimpong.

He looked at me, puzzled. "I can't do it." I said that it was all right and handed him a piece of paper, trying to reassure him that it really didn't matter what the map looked like. I told him that he didn't have to know *how* to draw a map, but perhaps he could just show me how he got from place to place, or simply mark down any landmarks he remembered. He looked uncomfortable. "It won't be correct," he said.

A Tibetan friend who was sitting with us told me quietly in English that Lobsang was probably hesitating because he didn't go to school, because he never learned how to make a proper map. "But that's the point!" I replied. "It won't be a *proper* map." I told her that Lobsang's map would eventually be one of several maps that I could use to represent traders' "alternative representations" of the region, or their "geographical imaginings," . . . or something like that. Confusion abounded. Lobsang shook his head. "I can't do it." It was soon clear that I had made him embarrassed. My face felt hot and I cringed, realizing that I had made some sort of naive ethnographic faux pas. I quickly switched the subject.

Asking traders to draw cognitive or mental maps was one part of the original methodology for this book. This request was, I thought, going to provide an unusual contribution to the field of anthropology; in addition to other frequently used methodologies such as participant observation, snowball sampling, and semistructured interviews, I planned to ask traders to provide rough drawings of their trade journeys, thus demonstrating how their spatial representations of the trade route differed from those depicted on "real" cartographic maps. After all, Barbara Aziz, one of the earliest anthropologists to research Tibetan socioeconomic issues, did the same with villagers from Tingri, in central Tibet, in the 1970s. She asked traders to create drawings of the region they inhabited, and they ended up drawing egocentric maps of villages and mountain peaks, depicting important economic hubs at the center, "spontaneously without any familiarity with modern mapping" (Aziz 1975: 32). Inspired by Aziz, I envisioned my project as an updated version of her efforts. By asking members of two generations of traders to draw such maps, I hoped to demonstrate how people's spatial conceptions of their trade journeys between Tibet and India had altered over time in relation to major changes in Sino-Indian infrastructure. Later, however, Lobsang's concern that his drawing would not fit what hegemonic maps *should* look like seemed perhaps more significant than any actual drawing of an "alternative" map, for only certain people even use maps in the first place.

In some cases, the creation of other kinds of maps can be useful. For instance, in a scene in *My Winnipeg*, filmmaker Guy Maddin's black-and-white "docu-fantasy" of his return to his wintry Canadian hometown, Maddin tells us that two competing taxi companies work the streets of Winnipeg. One uses the visible front lanes to get around. But because the streets are so snowy and unmanageable in winter, the other company takes passengers along the unnamed back lanes. Although no one talks about the alternative route in public, everybody in town knows about it. Accompanying the narration is an image of two street maps, one superimposed on the other. At some nodes the maps are at odds; at others they overlap and become blurry. Together they form the shape of the town, vividly demonstrating the importance of looking simultaneously at different ways of making or producing places: those that are hegemonic and seemingly more powerful (such as cartographic maps), along with others that might overlap or bump against them (such as the alternative, unspoken taxi route or, perhaps more interestingly, Lobsang's hesitation in making a map).

Attention paid to a certain new place or route (or one representation of a place or route) for hegemonic political and economic goals often results in the erasure, deflection, or obfuscation of other places. Such a "geographical diversion" also generates other kinds of diversions, which are often attempts to make the original place or route, or another place or route, visible or coherent again. Although the following two metaphors for diversions—a stream of water and a traffic sign—are relatively simple (and admittedly not wholly adequate), perhaps they may help elucidate how this process works.

Bear with me a little here. As a child builds a small dam of logs to redirect a stream, numerous things may happen: the water may back up, it can carve its way down the hill in several different directions, rocks may tumble from precipices and divert the stream further, the dam may backfire and completely overflow, or the stream may dry up altogether. Another analogy: diversions or detours are common in arenas of urban traffic. Let's say a government or municipality constructs a road connecting "Middletown" to "Worcester." This is one kind of diversion, deflecting attention from Middletown's relationship to other towns not along this path. And then during the construction, signs are erected to direct travelers to take another route—another diversion. Perhaps the alternative route is easy and direct for some residents, but perhaps others are taken much farther out of the way than they desire, and perhaps still others are stuck because their home is in the middle of the construction zone. Since travelers who live along these roads often know the local routes better than the municipal planners, they will sometimes devise alternate ways of getting where they need to go: through back streets, at different times to avoid traffic, or even by bicycle or on foot. These are additional diversions, and of course there are plenty of possibilities for others. Because the production of a geographical diversion—by the state, by groups who wish to reap economic profits, or by people simply going about their daily lives—brings some places more into view than others, the analogy provides an approach for examining the creation of favored places and the obscuring of others. It also gives one a sense of the varied kinds of mobility and immobility that can be generated by diversions: stopping and backing up, or using a path that takes a longer time, or cutting across different kinds of territory, or using alternate modes of transportation.

The trade route from Lhasa to Kalimpong cuts across a section of the eastern Himalayas that has been characterized by colossal economic and political shifts over the past five or six decades. Through the narratives of two generations of traders who have exchanged goods ranging from yak tails in the past to rice cookers and furniture now in Lhasa, Tibet; Kalimpong, India; and Kathmandu, Nepal, I elaborate on this notion of the "geographical diversion" with specific examples from my fieldwork, asking how such diversions are created, by whom, and for what reasons.

The book's main argument is as follows: diversions are produced by various actors, and it is the interplay between the apparent fixity of certain paths or boundaries and the mobility of local individuals around such restrictions that actually *produce* geographies and histories of trade. These geographies of trade are created from the level of the commodities the traders exchange, to the trade route itself, to the development of regions in Asia, and often—but not always—work against state notions of what the trade route *should* look like. There are also two subsidiary goals of this book. The first is to bring together narratives of people, places, and things in order to attempt to better connect the oft-divided micro and macro levels of research—in this case, to connect "everyday" trading experiences with abstract, larger-scale global economic processes. The second is to provide a platform from which to rethink capitalist processes in Asia (and perhaps even capitalist processes in general) by examining how geographical diversions are produced and enacted during various economic, political, technological, and historical transitions in the history of trade.

Those who have found that their towns or trading activities have become less accessible or less visible as a result of large-scale economic shifts often attempt to restore the coherence or visibility of their trading places. These tensions between mobility and fixity exist at every geographical level, from the nation-state in its attempt to curb immigration and petty trade in the name of "national security," to the trader who is restricted from carrying flour through one border crossing and moves to another; every scale generates its own dynamic histories of moving and stabilizing. By looking at trading practices that cross the boundaries of nation-states and ethnic groups, this book argues against binary "top-down versus bottom-up," "global versus local," or "formal versus informal" models of hegemony and resistance (where each is characterized by a certain kind of strategy such as large-scale powerful domination versus small-scale resistance), so that we may obtain a more nuanced picture of the tensions and overlaps between large-scale economic shifts and smaller-scale practices in the region. This book is ultimately a story about the collaborative and competing social practices that make a trade route a route, and a region a region.

Against Flows: Fixing the Route

Contemporary discourses of globalization and transnationalism are concerned with the rapid increase of people, capital, images, and commodities moving, migrating, or being transported over national borders, "compressing our sense of time and space and making the world feel smaller" (Inda and Rosaldo 2001: 4). Part of what emerged from some of the anthropological literature on globalization in the early 1990s was an emphasis on the increased mobility, "flows," and the border crossings of goods and people, as well as the notion that deterritorialization or the breaking down of borders was one consequence of this increase in movement (Appadurai 1991; Clifford 1997; Marcus 1995). While there is no doubt that capitalist development—which accelerated in the early 1990s, partly because of technological innovations and information networks—has led to the acceleration of migration and flows of capital in certain areas, I am more interested in highlighting the contradictory and uneven processes of globalization. This story is not so much about recent flows as about the *disruptions* of earlier flows, and these cross-border processes are not new, even in a region of the world like the Himalayan plateau, which is often depicted as "timeless" or "remote."

If globalization processes that take the form of escalating inequality, contradictions, and unevenness are characteristic of our current milieu, as critical geographers have acknowledged (Harvey 1990, 1999; Smith 1984), did these processes also characterize the past? How might we begin to explore tensions that exist between mobility and fixity, movement and stability over time? (I use the term *fixity* with the acute understanding that nothing is ever completely fixed or stable—even rocks and mountains wear down eventually; *anchoring* may be more accurate.) I call for exploring the nuanced relationships between barriers and movement instead of just focusing on the latter. Theorists of flows and movement have been criticized for "privileging theories of displacement over location" with the point that actual experiences of movement—in the case of refugees having to flee their homes, for example—are not always celebratory (Behdad 2005: 231). Moreover, at least one scholar has suggested that we ought to look more closely at differential mobility and how the *control of mobility* by some people weakens others, such as how the use of personal cars by some reduces others' ability to take public transport (Massey 1993: 62–63, my emphasis).

In this book, I angle my discussion of trade in the Himalayas toward "fixing" practices—both hegemonic and not—that might obstruct mobility for some. What I mean by "fixing" practices are those that attempt to make certain places more visible or coherent, both spatially and temporally. When a long-standing trade route is threatened to be rerouted or moved by political or economic intervention, various groups may find the need to make these places secure or visible again or to reinsert new meanings into older places that have been transformed or are no longer familiar.[2]

In the first chapter, which deals with the twentieth-century history of the Lhasa–Kalimpong route and British attempts to make it available for trade, the practices of "opening" and "closing" a border or a road are fraught with complexities that surpass simple understandings of the verbs *open* and *close*. Who is trying to halt others' access to mobility, for what reasons, with what consequences? I address this question not because I believe that stability or anchoring is more important than mobility but because these dynamics are in tension with each other, and ways of moving and fixing, or making places, need to be studied together. In a late capitalist world characterized by "flows," it is important to ask where we can locate the tensions between geographical fixity and mobility; that is, which areas or people are becoming more mobile and which areas or people are becoming more stable? Are these differences related to generations, class, or gender, for instance? To further complicate the question, geographical transformations are produced not only at the level of the state but all the way down to the level of the individual trader, and even, as I argue in this book, at the level of the object.

One useful yet abstract way of theorizing the "fixing of place" from the perspective of large-scale histories of capital can be found in David Harvey's work, where he discusses two kinds of "fixing" strategies: the idea of "fixed capital" and the "spatio-temporal fix." Fixed capital is a physical—but not necessarily immobile—form of capital (such as caravan mules, a train, or an airport), which is "literally fixed in and on the land in some physical form for a relatively long period of time" (Harvey 2003: 115). In fact, Marx, in volume 2 of *Capital*, warns his readers not to assume that immobile objects are automatically fixed capital (Marx 1992: 242, 288–89). In other words, economic fixity is not the same as physical or geographical fixity. What is important for our purposes here is that investments in "fixed" things (whether in yaks, a road, or a credit system) are made in order to allow goods or services to move more quickly and more profits to be reaped in the future. Fixed capital, however, can also act as a spatio-temporal "fix" to a problem: "a particular kind of solution to capitalist crises through temporal deferral and geographical expansion" (Harvey 2003: 115).

One of the ironies of the movement of capital is that it creates its own barriers to further accumulation (Mitchell 1996: 112). The following chapters question how large-scale capitalist practices of fixing place—and especially refixing place—are both experienced and enacted by traders in Asia. The chapters also examine how these "refixing" strategies can work both against and with larger-scale transformations. As mentioned earlier, one of the goals here is to contribute to critical debates in the social sciences that seek to connect the often abstract rhetoric of globalization and transnationalism with its more local experiences. In doing so, I hope also to extend arguments against state-centered notions of globalization or "national economies" (van Schendel 2002; Winichakul 1994). But first, in order to situate this study in its regional and historical context, I present a brief outline of the social history of the Lhasa–Kalimpong route.

Shifting Economic Geographies in the Himalayan Region

In contrast to early Western cartographers' imaginings of Tibet as an inaccessible "blank" on the map between China and India—one to be filled in by explorers—a web of extensive trade and pilgrimage routes have cut across the Himalayan region since at least the seventh century CE (Bishop 1989: 75; Madan 2004). Asian travelers, politicians, traders, and religious figures recorded detailed descriptions of the commodities Tibet had to offer; the region was particularly known for its musk, furs, wool, borax, and gold (Boulnois 2003: 133–56; Klimburg 1982: 25–37; Schafer 1985). The Barkor, the central pilgrimage circuit around the holiest temple in Lhasa, was (and still is, to an extent) a bustling site of exchange for a variety of goods—chunks of lapis from Afghanistan, raisins from Khotan, brick tea from Sichuan, and cotton from India—which can often stand as material indicators of the long-standing economic connections between Lhasa and other parts of Asia.

Although nearly all of the trade routes between Tibet and other parts of Asia have been established, cut off, and reestablished over the centuries as a result of

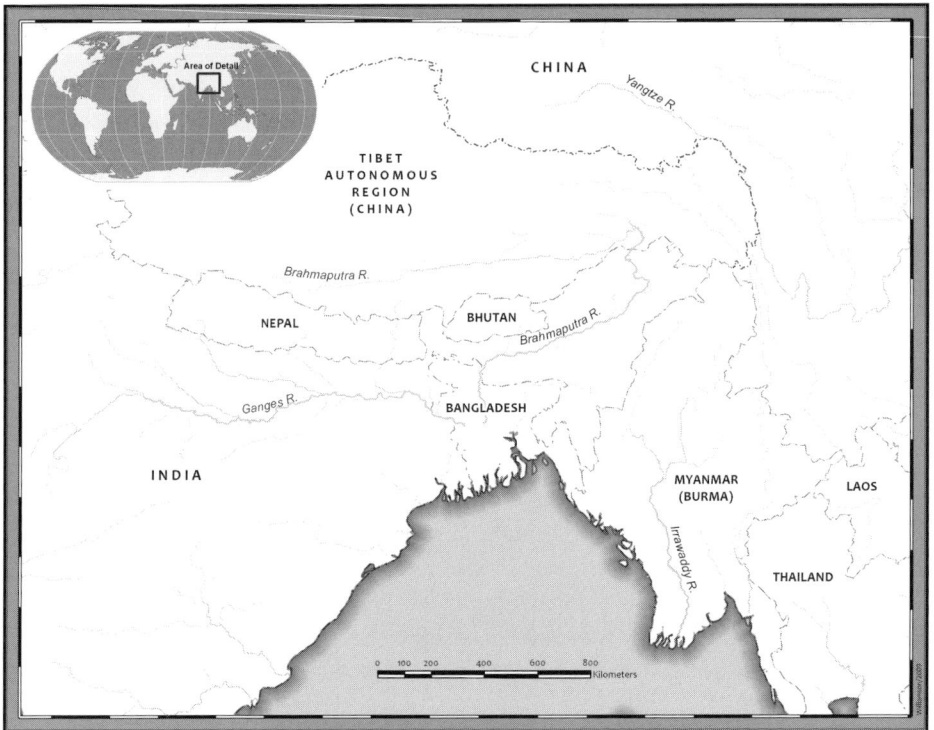

FIGURE 1. The eastern Himalayan region. Doug Williamson, Hunter College Department of Geography.

wars or economic necessity, this story mainly concerns the Lhasa–Kalimpong route and some of its farther-reaching geographical nodes, including Kolkata (Calcutta), Guangzhou, and Singapore.[3] The expansion of capital networks in tandem with designs on territory is a common thread throughout the history of the region. For the purposes of this book, we begin in the mid-1800s, when claims on Tibet were partly a result of the desire for new markets after the economic decline of the East India Company and partly a result of an interest in strategic expansion; in effect, the British hoped that their goods could be introduced to China through the "back door" of Tibet.[4] The 1903–4 invasion of Tibet led by the British commissioner to Tibet, Francis Younghusband, resulted in a number of events that reconfigured the landscape: the "opening" of the Chumbi Valley between Sikkim and Tibet for trade (despite the fact that small-scale trade had continued over those very mountain ranges for centuries), numerous road-building enterprises, and the establishment of British Trade Agencies in Gyantse, Yatung (*gro mo* in Tibetan), and Lhasa. These institutions, osten-

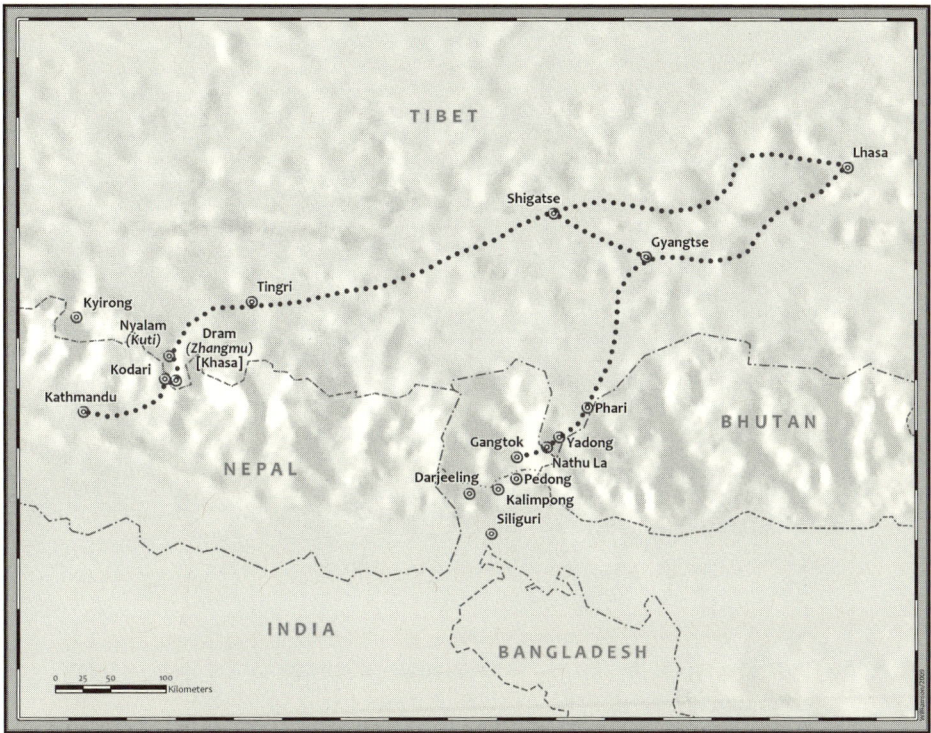

FIGURE 2. Two of the routes currently used for trade between Tibet, India, and Nepal. Doug Williamson, Hunter College Department of Geography.

sibly set up to manage commerce in the region, were essentially created and maintained to prevent the possibility of Russian, Chinese, and later Japanese encroachment on British Indian territory.

With the Chumbi Valley controlled by the British, the Jelep-la mountain pass on the border of Tibet and Sikkim was "opened," linking Lhasa, Gyantse, and Phari (see figure 2) with Kalimpong. Roads were also built up to the Nathu-la pass (the route that linked Lhasa with Gangtok), which at the time saw less traffic than Jelep-la because of its more treacherous topography. Before long, Kalimpong, originally a small town marked by local trade with Bhutan and Tibet, grew rapidly; it soon became a major center for British-run missionary schools and colonial holiday goers. Meanwhile, the Kathmandu–Lhasa trade routes through Kodari and Kuti (see figure 2) were cut off, superseded by the shorter route through Jelep-la (see figure 4). As a result, many merchants and traders from Nepal also settled in Kalimpong.

By the mid-twentieth century, Kalimpong had become one of the most im-

FIGURE 3. The Barkor in Lhasa, with shops and vendor stalls, 2002. Tina Harris.

portant towns in the region for the exchange of commodities; in particular, it was the main center for the sorting and processing of Tibetan wool. The wool was brought to Lhasa from the Ngari and Changtang areas of western and northern Tibet by nomads and middlemen. In Lhasa, the wool was baled for transport and loaded on mule caravans, led almost exclusively by Tibetan or Newar (a cultural group from the Kathmandu Valley) men, indicative of a gendered division of labor in trade. Men tended to be mobile, leading the mule caravans from town to town between Lhasa and Kalimpong, whereas women would often stay behind, managing the shops or stalls and occasionally the finances. Gendered divisions like this were (and are) also subject to tensions between mobility and fixity; in order for the muleteers and wool to move, there had to be a foundation or mooring from which this mobility could be based.

The caravans would make the approximately one-month-long journey in stages, stopping in Gyantse, Phari, and Yatung, traversing Jelep-la, and eventually ending up in Kalimpong. Here the wool would be weighed, sorted, and stored in large *godowns* (warehouses) for transport to the port of Kolkata. Groups of mules would return to Lhasa laden with Indian grains, household implements, cotton garments, and even Rolex watches. Due to growing international demand, wool prices were at their peak in the 1940s. During the period from April 1, 1946, to March 31, 1947, the export of wool from Tibet to

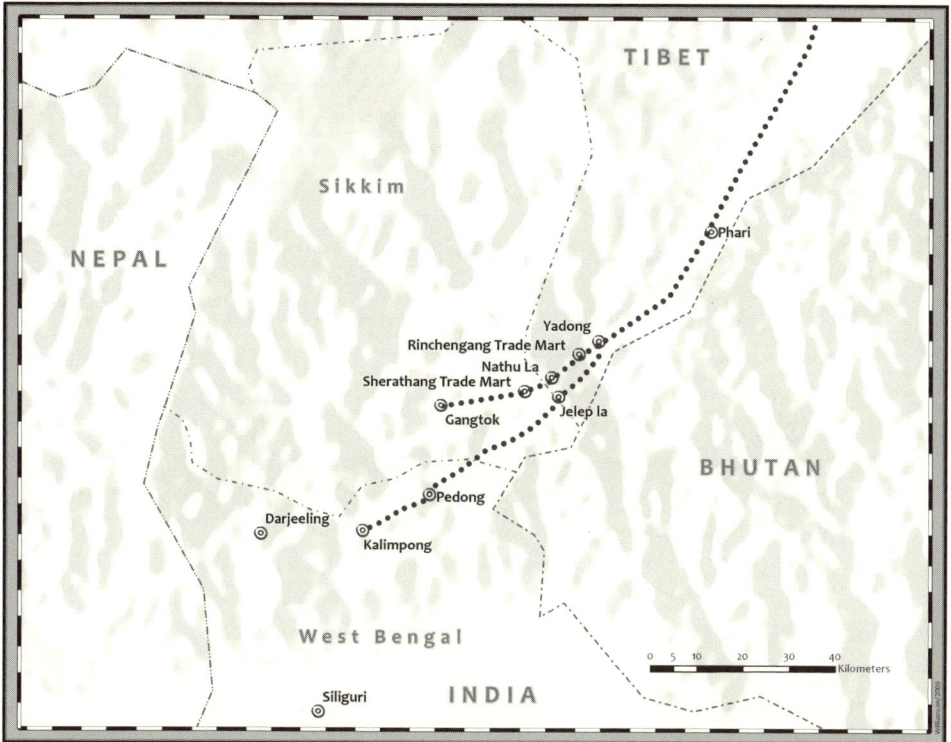

FIGURE 4. Location of Nathu-la (reopened in 2006) and Jelep-la (closed since 1962). Doug Williamson, Hunter College Department of Geography.

Kalimpong amounted to 106,615 maunds (approximately 8,742,430 pounds), at 55 rupees per maund (then about US$1,766,212). At this time, the three largest trading families in Tibet—Pangdatsang, Sadhutsang, and Reting—held a monopoly over the wool trade, while some smaller-scale traders did not fare as well. In 1945, for example, "the petty traders could not hold their wool for various reasons and they had to dispose of their wool to the big traders at comparatively low prices" (India Office Records, 1945).

According to one man who lived in Lhasa during the 1940s and 1950s, the economic connections between Kalimpong and Tibet were so important to the local geographical imagination that many Tibetans would use the word "Kalimpong" to refer to India as a whole. It is this position of Kalimpong in the world of twentieth-century Himalayan trade networks that still seeps up through contemporary accounts of trade. Along with the picture of Kalimpong as a cosmopolitan trading hub came the remembrances of languages; how

Hindi, Marwari, Nepali, Newari, Chinese, English, and Tibetan were common languages spoken in the city, but Tibetan was truly the language of trade. One Marwari trader recalls that no Tibetan really knew Indian languages when it came to the wool trade, but there were certainly many Indian businessmen in Kalimpong who could speak excellent Tibetan. "If you went to India, you did not need a translator."

After the Chinese People's Liberation Army (PLA) occupied Tibet in 1950, the trade in wool began to decline considerably. Tsering Shakya has noted that by the 1950s, 70 percent of trade between Tibet and India was effectively bought out by the Chinese State Trading Company, cutting off the businesses of long-established Newar and Marwari traders (business families originally from Rajasthan) in Lhasa (Shakya 1999: 115). At the same time, however, the presence of the Chinese army in Tibet was a boon for some traders, who quickly switched from selling wool to selling more lucrative commodities that were in demand by the soldiers, such as batteries and petrol. In 1962, however, boundary disputes and the Sino-Indian War resulted in the sealing-off of the border passes between China and India. Although this study does not address the lengthy and complex geopolitical background of Sino-Indian border conflicts, the main areas of contention that led up to the 1962 war were disputes over the Aksai Chin territory between Ladakh, Tibet, and Xinjiang and the McMahon Line between Arunachal Pradesh and Tibet. It is noteworthy that when Indian boundary lines were drawn up by the British in the late nineteenth century, several "no-man's land" areas—although checkered by small villages and inhabited by numerous tribes—were drawn on maps at the last minute and left "open to modification" (Maxwell 1970: 61). Now, a couple of hundred years later, most commercial maps of the region sold in India are stamped in blurry ink with the sentence, "The external boundaries of India as depicted are neither accurate nor authentic."

After 1962, then, trade along the Lhasa–Kalimpong routes was almost completely cut off, significantly transforming the economic geography of the region. Many traders, some of whom were given only twenty-four hours' notice to vacate their shops in the border marts, shifted their businesses by settling in Kathmandu and rerouting their trading links with Tibet through Nepal. Since the Newars were originally from Kathmandu, it was slightly easier for some of these families to create these trading links. This new route was the next-closest officially open route to Lhasa, passing from Kathmandu through the border checkpoint at Kodari on the Nepal side, then through Dram (Khasa in Nepali, Zhangmu in Chinese) in Tibet, and on through Gyantse to Lhasa (see figure 2). By 1965, a paved road between Kathmandu and Kodari—as well as the Friendship Bridge linking Nepal and China—became fully operational, transforming the Kathmandu–Lhasa route into the main route for Nepal–Tibet overland trade (and later tourism).

After the start of Deng Xiaoping's reforms in 1978, state control over trade in Tibet began to diminish, and the subsequent growth of China and India as major economic powers led to renewed Sino-Indian trade negotiations in the 1990s and first few years of the twenty-first century.[5] In 2003, concurrent with rapid increases in development in both China and India, Indian prime minister Atal Bihari Vajpayee visited Beijing and signed a declaration with Chinese premier Wen Jiabao to expand border trade at Nathu-la. This document, while emphasizing the opening of Sino-Indian borders for the "free" movement of commodities, was simultaneously an agreement wherein India accepted the Tibet Autonomous Region as an integral, bounded part of the People's Republic of China. Whether or not the signing of this document meant that China implicitly accepted Sikkim as a part of India remains debatable. (Sikkim went from a protectorate to a state of India in 1975.)

The research for this book, conducted between 2005 and 2007, took place at a time when a surge of major state-led development projects in Tibet and other western regions of China, such as the Qinghai–Tibet railway, were well under way or completed. In 2006, amid much fanfare, the Nathu-la pass between India and Tibet reopened for trade. At the time of research, many traders, most of whom were children or grandchildren of former Kalimpong–Lhasa traders from the 1940s and 1950s, were actively looking for opportunities to sell inexpensive and profitable Chinese goods directly through southwestern port cities like Guangzhou and the Special Economic Zone of Shenzhen, which had quickly become major national and global manufacturing sites. At the same time, other traders who sold local Indian, Tibetan, or Nepali products or commodities found themselves struggling to compete, and several of the older generation of traders I interviewed for this book and who remembered the pre-1960s trade had unfortunately passed away. It is against this backdrop of significant social, economic, and geographical changes that the drama unfolds.

The People and the Places: Tibet, the Himalayas, Ethnicity, and Traders

There have been many, often contradictory variations in the production and interpretation of images of the Himalayan landscape. Western, Chinese, Indian, and Tibetan perspectives found in travel writings, anthropological work, and tourist accounts have shaped and reshaped the region as a mystical "Shangri-la," as a territory ripe for mineral exploitation and urban development, as a feudal theocracy, as a borderland marred by crime and prostitution, and so forth. In addition, the contemporary study of Tibet and Tibetan-speaking communities in the Himalayas cannot be detached from Tibet's complex politico-religious history. Many political and cultural histories of the region have focused on the debate over whether Tibet has historically always been part of China or an independent entity, or over which particular moments in Tibetan history have been characterized by autonomy and which by subjugation. And although the influence and transformation of Buddhist practices in Tibetan-speaking Himalayan communities has until recently remained a dominant theme in work concerning the area, this book centers instead on the contemporary economic history and shifting geographies of Tibet and the Himalayas.

Such a focus is intended to suggest that the struggle over the control of trade routes and economic networks has been a crucial factor in the production of the history and geography of the region. To take a few widely separated examples, the development of eighth-century Silk Road trade routes to Lhasa facilitated the passage of religious tribute caravans; the British invasion of Tibet in 1904 was partly prompted by Britain's declining control over trade and its colonial properties; and the Tibetan Trade Delegation, formed around the time of Indian independence in 1947, served to promote the idea of Tibet as an independent nation while at the same time ostensibly exploring trade agreements with India, China, and the United States (Cammann 1970; Klimburg 1982; Shakabpa 1967).

Much academic writing concerning Tibet or Tibetans has until recently consisted of fieldwork among Tibetan populations in exile communities in Nepal and India, studies of Western representations of Tibet and Tibetan Buddhism, or research outside the city of Lhasa (see, for example, Bishop 1989; Diehl 2002; Dodin and Räther 2001; Goldstein and Beall 1989; Klieger 1992; Makley 2003; McGuckin 1997). With the relaxations of visa requirements for researchers who wished to work in the TAR between 2001 and 2007, as well as the establishment of official partnerships between foreign universities and academic institutions in Tibet between 2000 and 2008, several excellent texts on contemporary Tibetan medicine, modern art, historical politics, and agriculture began to emerge during this time period (see, for example, V. Adams 2002; Barnett 2006; Yeh 2003).[6]

While I began this project intending to interview "Tibetan traders," it soon became clear to me that I could not concentrate on a single ethnic community, so I departed from the more common method of centering on a specific ethnic or cultural cross-border trading community ("Tibetan traders" or "Marwari merchants"). The wool trade from Lhasa to Kalimpong, for instance, involves complex networks of nomads, intermediaries, and merchants that transcend territory and, accordingly, any sole ethnic category. Sheep, for example, might be herded by nomadic Amdowas (a Tibetan-speaking regional group from the northeast); the wool might be purchased and transported to Lhasa by the Hui (Chinese Muslims), carded and spun by half-Tibetan half-Nepali women, brought over the mountain pass by Newar muleteers (a Buddhist and Hindu group from Kathmandu, Nepal), and sold in Sikkim by Marwari merchants (business families originally from Rajasthan) or by Bhutias (Tibetans who have lived in Sikkim for centuries).

The trade between Kalimpong and Lhasa has never been restricted exclusively to one ethnic group. Prior to 1962, Marwari moneylenders, elite Tibetan politicians, Newar merchants, and Tibetan Muslim muleteers, although continuing to maintain their community identities in various ways, encountered each other daily within a similar social and economic sphere. For instance, in a historical memoir of Newar trade, one man living in Kalimpong had a shop that "became the meeting place for the Newaris, Tibetans, and Marwaris: the place to get new information. In particular, he put in a good word for other traders so they could become prospective business partners; he became someone all could trust, and soon became the most influential middleman in Kalimpong" (Hilker 2005: 42–43). Today, although the ethnic and class backgrounds of trading communities along the route may have altered slightly, diversity is still important; varied stakeholders remain jointly engaged in struggles over the new economic realities of the region. It is these types of transnational interactions—regardless of whether the various groups actually "got along"—that led me to use as a framework for this project the Lhasa–Kalimpong trade route rather than ethnic groups, the study of which (even since Eric Wolf's critique against treating cultural groups as "billiard balls") can leave the impression that they exist in isolation or as self-contained cultural units.

But this is not to say that these groups failed to uphold a level of community or cultural coherence; for instance, the Tuladhars, a Newar family that owned shops and offices in Kathmandu, Lhasa, and Kolkata, would send male family members to Tibet for extended periods of time, between three and ten years, from as early as the mid-1800s until the 1950s, while the women would remain at home in Kathmandu (Tuladhar 2004: 18–19). The merchants in Lhasa established community centers, organized festivals, and maintained *guthis*, "socio-religious organizations that unite individuals to perform certain tasks which often have a religious goal or motive at base" (Lall 2001; Lewis 2003: 51; Tuladhar 2004: 20–21). Nevertheless, minority groups do not necessarily share common experiences of oppression and "a common goal to displace hegemonic powers" vis-à-vis "the majority" (Lionnet and Shih 2005: 234). This is true, moreover, not just between different minority groups, but even within the category of "Tibetans." Since the time of the Dalai Lama's departure from Tibet in 1959, a number of Tibetans have established active exile communities in India and Nepal. These communities have developed along considerably different social and political lines from the majority of Tibetans who still reside in the TAR, as well as from those who live in other Tibetan-speaking communities in China. Yet there remains a tendency for scholars to base their knowledge of "Tibetans" on research in a single community. Though this tendency partly results from heavy restrictions on research in the Himalayan region, it need not be a given. I hope I might provide a bit of a corrective by highlighting the cultural and political specificities of the region's varied communities while at the same time acknowledging the importance of transnational social and economic networks, the territorial and economic struggles taking place in communities on both sides of the Sino-Indian border, and the politically charged meanings of the term *nationalist*, sometimes used as synonymous with *separatist* in the TAR. I also hope to introduce an important new trajectory: the development of a transnational entrepreneurial elite who are neither full-blown advocates of a Tibetan independence movement nor staunch supporters of the Chinese regime. Studies of the development of this new "modern" elite, as well as research that investigates the creation of official ethnic minority categories in the PRC, raise issues such as the authenticity of Tibetan-ness (versus Chinese-ness) and how urban communities are divided along ethnic and class lines.

What do I mean, then, by *trader*? In earlier centuries, "the word 'trade' meant a craft, to distinguish it from a commercial calling," but around the time British imperialism reached its zenith in the nineteenth century, it became "associated with commerce and risk taking . . . the word [had] swung around completely" (Adams 1974: 141). The role of the "trader"—at least in the West—shifted from one of production to one of exchange and profit. In much of the existing literature on trade, there is a conscious effort to distinguish merchants—specialists in mediation for profit—from traders, whose roles could include producing the goods they exchanged (R. Adams 1974: 239). John Middleton, in his study of Swahili merchants, has distinguished between the roles of merchants and traders by stating that "merchants differ from traders in that they act as intermediaries between different producers and consumers," whereas the role of traders, who tend to be more mobile, is limited to the direct exchange of what they have produced with what someone else has produced (Harris 2007; Middleton 2003: 509–10). Indeed, I have seen many traders acting as producers or vice versa; for example, a small-scale trader might make woolen jackets in his home village and then bring them to the Lhasan marketplace to sell. However, in my experience, the line between merchants and traders often seems fuzzy; both may act as intermediaries and negotiators who bridge the distance between production and consumption by participating in local exchange (Steiner 1994: 130). In fact, the Tibetan term for "trader," *tshong pa* or *tshong mkhan*, usually implies both "trader" and "merchant," as does the Nepali word, *"byāpāri."* According to some people I spoke with, being called a trader is sometimes considered derogatory; the occupation has been historically associated with cheating and swindling, and certain professions, such as butchering or dealing in leather shoes, are occasionally still looked down upon, as they were often linked to non-Buddhists (who are allowed to kill animals) or lower castes or classes (Braudel 1992: 558; Humphrey 1999). As one successful Lhasa-based trader told me (in English):

> There are also some sorts of rumors about businessmen. Not bad rumors, but what do you call them—stereotypes—about businessmen. A businessman implies that you are wise, and sometimes this is negative. Dishonorable, this type of stuff. There are also a lot of traditions and perceptions involved in the culture as well as a businessman, with various judgments as far as religion goes, like meat traders, and leather makers. These kinds of things are actually looked down upon, and they are not valued as a sort of occupation. But, they are obviously business-people and are very actively involved in social aspects. But here, they are looked down, and these people are sometimes segregated out here. But trading happens, regardless of who you are! You gotta eat meat, so you still have to go out and find these people.

In the context of this study, then, I consider traders to be "middlemen of varying degrees," although I do not make further distinctions between different types of middlemen or intermediaries, as Caroline Humphrey does in her work on post-socialist trade and barter in Russia (e.g., "shuttlers," who travel with their goods, and "trader-retailers," who have the money to purchase a small shop in addition to their trading activities; Humphrey 1999: 37–38). I am more concerned with how the stratified "degrees" of trading specialization or identity may emerge from stories and experiences along the trade route and embody tensions between mobility and fixity. As we shall see, there are trading elites, such as the three wealthy Tibetan families from the Kham region of Tibet who held the monopoly over wool. By contrast, traders who bring Chinese toilet paper to local shops in Nepal dramatize the differences between "big" and "small" traders. There are stratifications in gender: women remain at retail shops in town while the men move with the mule caravans. The gap between different kinds of trading groups seems to be increasing as unequal economic and geographic development also escalates. Finally, although some trading practices may occasionally take the shape of "everyday forms of resistance" or "tactics," others at times overlap with and contribute to powerful state practices. The point is to show that various kinds of trade routes are in fact *created* by all these kinds of practices.

Cartographic Anxiety

I present this general historical, geographical, and cultural outline of the shifts in trade, politics, and people between Lhasa and Kalimpong for two related reasons: first, to give readers a basic sense of the seismic political and economic shifts in the region over a period of approximately sixty years; and second, to demonstrate that tensions between mobility and fixity and the creation of diversions on numerous geographical scales have been common features over the course of the economic history of the region. For instance, the road that was built by the British in order to facilitate the movement of goods and people through the Chumbi Valley in 1904 simultaneously cut off trade routes in other parts of the region; when the wool trade declined and the border between Sikkim and Tibet closed in 1962, traders turned to more profitable commodities and moved their businesses to more lucrative locations. Through these actions we can begin to see how the region and representations of the region are produced and understood by competing groups today; past events tied to the landscape play a part in the creation of the region, not only "on the ground," so to speak, but also in the contemporary geographical imagination.

Territory is never stationary, and yet competing groups will often attempt to make the land more visible or coherent if it becomes threatened or occluded. Certain kinds of images or representations of the region's history can be reshaped by creating or recalling alternative histories of space. For instance, when some elderly traders talk about the "golden era of Lhasa–Kalimpong trade," their narratives contribute to making a particular representation of the region, located in a very specific time and place. Memories of the trade route as part of the "golden era" of the early twentieth century are directly set against how the present or the future state of the economy is perceived; in this case, for example, traders worry that the recent influx of inexpensive Chinese commodities will effectively outsell locally made goods such as incense.

The scene with Lobsang at the beginning of this introduction was not something I thought too much about during my time in the field (in fact, it hardly even made it as an entry in my field notes), but memories of the incident began bothering me when I began to make sense of my notes and journal entries. One of the questions that kept recurring was, why couldn't Lobsang make the map? There is, of course, no "correct" answer. He may have been worried about surveillance; border regions are sensitive, and talking about the geography of the area or making and utilizing maps is something that only officials or authorities do. The TAR is not an easy place for cultivating the high level of mutual trust that is helpful in ethnographic research; perhaps he was worried that he could get into trouble with the authorities by talking about the area in detail to a foreigner. Or perhaps my friend was right; given my position as a scholar, Lobsang could have felt that I was quizzing him, exposing some kind of ignorance. Maybe he simply didn't want to do it. But if we take Lobsang's reaction at face value, perhaps there is another, more obvious answer: the map won't be correct, because it simply *isn't* correct. In other words, the abstractness of any cartographic map—or indeed any visual representation of Lobsang's trade journey—will never match up with the actual lived experience of trading.

As Neil Smith has written, "consciousness of space is a direct efflux of practical activity"; hence the gap between practical physical activity and the cartographic representation of abstract "space in general" is enormous and laden with power imbalances (Smith 1984: 96). Established modern maps, "in which a kilometer is a kilometer no matter what the terrain or body of water, are therefore profoundly misleading in this respect" (Scott 2009: 47). James Scott's explanation here is poignant—allow me to paraphrase. If you ask a peasant living in the hills of Burma how far it is to the next village, the answer will probably be in units of time, not in terms of distance. The reason is this: Place A may only be twenty kilometers from Place B. But depending on the difficulty of travel across flooded or mountainous terrain, not to mention the kind of transport used, it could be a one-day trip or a five-day trip. Furthermore, if Place B is located lower in a valley and Place A is high up in the mountains, the uphill trip from B to A will certainly be longer and more arduous than the downhill trip from A to B, though the actual distance in kilometers is the same. Paying careful attention to this very "friction of terrain," or "friction of distance," "allows societies, cultural zones, and even states that would otherwise be obscured by abstract distance to spring suddenly into view" (Scott 2009: 48).

Further, "trade route" in the singular is a misnomer. It is not, as we might imagine it, one unbroken line on a map leading from Town A to Town B. In the everyday experience of trading goods from one town to another, there is never a single, smooth path. There are blizzards and detours; there are different tributaries that branch off; there are temporal pauses when a trader must stay in one town for three months or more; there are invasive passport checks; there are multiple stops and starts. Money may travel in one direction, people in another, and goods in yet another. There are, in effect, many varieties of diversions. Town A and Town B, therefore, do not stand isolated on their own, but are recognized as part of a long-standing web of social and economic networks stretching far beyond their city limits and the current milieu. Lobsang's reaction is telling because it highlights, even in its hesitation or refusal, the hegemony of a certain abstract way of fixing place, separated from the reality of lived experience. It is reminiscent of the idea of "cartographic anxiety," an anxiety that stems from knowing that a map is only a representation of a supposed objectivity, and the resulting sense of the separation between "the knower and the world" (Pickles 2004: 195, n. 4).

By looking at the competing processes of making places more visible or coherent, this research is intended to contribute to transnational studies and border studies, foregrounding an explicit conversation between anthropology and geography. Mapping, with its hegemonic history, is simply one way of fixing a place or places. Making or fixing a trade route can take the form of simply saying that a trade object comes from Tibet and not China or Nepal. It can take the form of going to one trading town and bypassing another, perhaps because of a landslide or a known security block. It can take the form of images and representations of trade hubs remembered from narratives of past generations. In British colonial accounts, the trade route between Lhasa and Kalimpong was "a region to be manipulated or overcome in the individual's search for a given destination, not an area to be lived in and through" (Black 2000: 13). The goal in this book is to try to discover why such geographical diversions are formed, who created them, and for what particular interests, and to show by contrast how the route is in fact "*lived in and through.*"

Chapter Organization

The goal of this initial chapter was twofold: first, to give readers a basic sense of the major political and economic shifts in the Sino-Indian region over a period of approximately sixty years, and second, to demonstrate that tensions between mobility and fixity and the creation of diversions on numerous geographical scales have been common features over the course of the history of the region.

Chapter 1 lays out the historical background to the current political, social, and economic climate along the trade routes and is supplemented by ethnographic accounts from older traders who were operating prior to and just after the closing of Nathu-la pass in 1962. It focuses on several snapshots of specific moments in the history of the trade routes that are significant to the rest of the book, moments that raise questions about competing spatial representations, social mobility, and the strategic directionality of trade. These snapshots take place during roughly three periods: first, the British mapping of the Himalayas between the late eighteenth and early twentieth centuries in order to open the Chumbi Valley in Tibet for trade; second, the peak and decline of the Tibetan wool trade; and third, the introduction of new kinds of transport in the mid-twentieth century. Finally, I examine the rerouting of the Tibet–India trade through Nepal after the 1962 Sino-Indian border war, as well as a ghost story that appears to link border disputes from the past with the more contemporary drive to increase Sino-Indian border trade over the newly reopened route.

I narrow the focus in chapter 2 to concentrate on traders' experiences with one specific commodity, wool. Since sheep wool has historically been the main item of trade in this region, I use interviews with those who were and are involved in wool production and trading networks (including the production of carpets or woolen handicrafts for the Western market) to see where the networks intersect with newer networks, such as the sale of rayon thread from eastern China. Personal narratives of the properties of commodities, as well as their spatial trajectories and origins, can serve as examples of "fixing" or making visible various kinds of geographies of place against others. This analysis follows the path of the *pang gdan*, a deceptively mundane Tibetan women's woolen apron, to show how the attachment of commodities to representations of place figures importantly both in the contemporary study of material culture and in uneven development. In the section "Narratives of Decline," I examine traders' descriptions of the contemporary decline of the quality of commodities and of trade, demonstrating that values of purity and life or dirt and death are attached to goods in very specific times and locales and acquire particular meanings in these contexts. Toward the end of the chapter, I return to the narratives of purity and decline by exploring what individuals are currently doing to revitalize the trade in Tibetan wool for local and international markets.

The third chapter begins to look at how and why some places in the region become favored over others as a result of state-level infrastructural diversions. I examine several discourses over the reopening of the trade route that historically connected Lhasa in the Tibet Autonomous Region (TAR) with Kalimpong in West Bengal, demonstrating that border regions are critical sites for highlighting the tensions between spatial and temporal mobility and fixity. I start with a theoretical overview that stresses the importance of looking at transnational mobility across border regions and argue that in addition, a closer examination of how various groups "fix" or make places more coherent through narrative is vital to the understanding of how economic change in Tibet, India, and China is experienced. Then, in order to illustrate how new spatial representations of the region are created and anchored in geographical imaginings of the past, I follow with three ethnographic stories. First, I look at contending spatial and temporal discourses over the naming of the reopened Nathu-la trade route as the "Silk Road." Second, I discuss how two groups—one for and one against the reopening of Jelep-la, an additional pass—both coincide with and struggle against state-centered conceptions of space. Finally, I examine how concerns regarding the geographical marginalization of Kalimpong (as a result of the reopening of the Nathu-la pass in 2006) are provoking local businesspeople to make the town more visible through the marketing of local products.

The fourth chapter continues the book's ethnographic inquiry into the tensions between mobility and fixity, this time at the national level, outlining some of the dynamics between state-level regulations and traders' interpretations of (and occasionally movements against) these regulations. I begin with a brief outline of some of the major claims made in the field of border studies, as well as some departures I make from this literature. Barriers are fixed to mark and securitize national territory, yet such state-based stabilization processes often go hand in hand with the notion of economic prosperity based on "free and open flows" of global trade. I argue that "opening" a border or "freeing" new spaces for trade paradoxically provides more opportunities for regulation. I then turn to specific narratives of trade practices across Chinese, Nepali, and Indian borders and attempt to demonstrate how these practices create trading places and routes that are integral to the history of the region. The case studies in this chapter include traders' experiences with new ID laws (and how they work around these laws), the kinds of commodities they are allowed to trade, and how and why certain commodities are sometimes turned into gifts. Here, I also briefly address the (il)licit border trade in the Himalayan region, stressing that traders do not always resist state rhetoric but will collaborate with state regulations at specific times in order to make a profit. Toward the end of the chapter, I zoom back out to a regional scale, showing how the practices just described are manifested geographically and reflect increasing economic and social gaps between "small" and "big" traders. Ultimately, the goal of this chapter is to show that borders are represented by the state in one way but actually *lived* in others.

Chapter 5 describes how the Himalayan region has changed economically and geographically over the past sixty years and attempts to fit its economic history into the larger story of long-term global economic shifts. As China is extending its capital development westward into Tibet and searching for new markets in South Asia, cities such as Kalimpong and Kathmandu are being by-passed. By looking to the Qinghai–Tibet railroad and new border openings as opportunities for new market niches, as well as by examining the global handicrafts market, I show how entrepreneurs throughout the region are forg-ing new trading networks that reveal a different topography than that of their grandparents' generation. Here, I explore accounts from three different groups of younger traders, showing how such directional shifts in trade have simulta-neously given rise to seemingly "successful" and "unsuccessful" experiences of exchange. While some traders' new paths have resulted in the establishment of social connections in far-flung cities such as Singapore and New York, the faster communication of trade information, and the diversification of trade commod-ities, other traders have found themselves in the exact opposite situation, need-ing to halt, slow down, scale back, or reverse their trading activities (such as reverting to barter) in order to survive. I propose that David Harvey's analysis of time-space compression is a model that does not fully reflect these divergent reversals of trade, and I present these cases of unevenness as further examples of how spatial solutions to capital flows can create new problems.

The final chapter revisits the notion of geographical diversions in conjunc-tion with Henri Lefebvre's notion of ideological "blind fields" and suggests how this particular study of hegemonic economic geographies can contribute to a more nuanced understanding of capitalist processes in Asia. I make two spe-cific considerations aimed toward those who study globalization processes in general. First, since spatial expressions of fixity have been relatively neglected by scholars of globalization in favor of tropes of mobility, I suggest that future studies pay closer ethnographic attention to the ways in which practices of fix-ing or making places more coherent are inherent to both capitalism's survival and struggles against dispossession. Second, I argue that it is extremely im-portant to examine individuals' lived experiences and their responses to major spatial transformations, since major sociohistorical shifts—especially in this case—are not always tied to specific modes of production. Shifts in large-scale economic geographies have less to do with capitalism or communism or any abstract changes in modes of production than they do with concrete events—in this case, border closings and openings, the introduction of new commodities, and the emergence of new kinds of infrastructure and modes of transport. I suggest that scholars of contemporary China and India often attempt to char-acterize this supposedly new kind of Asia-centered globalization by specific modes of production without paying attention to local narratives of place, and by doing so, become stuck in ideological "blind fields."

A Word on Methodology and Exposition

How can the focus of this book be simultaneously about people, places, *and* things? Isn't it easier and more effective to conduct an in-depth study of one of the elements that make up the story of the Lhasa–Kalimpong trade route, such as a city (e.g., Kalimpong), or an ethnic group (e.g., Tibetans), or an object (e.g., wool), or a theme (e.g., globalization)? After all, George Marcus described these methodological trends (which were relatively "new" in 1995) in his oft-cited *Annual Review of Anthropology* article on multi-sited ethnography. Marcus noted that large-scale concepts in globalization theory—such as world systems or flexible accumulation—were not addressed very well within single-sited ethnography, and that more interdisciplinary and cross-regional methods that followed the paths of people, places, or things were better suited to anthropological research in a rapidly globalizing world (Marcus 1995: 98).

Inspired both by Marcus's work and by the essays in Arjun Appadurai's 1986 *The Social Life of Things*—particularly Igor Kopytoff's essay on tracing the "biography" of an object—I initially thought that this project was one where I could "follow the thing" from consumption to production, along an established trade route (Marcus 1995: 98). I planned to choose a commodity from the main marketplace in Kalimpong (say, a Tibetan wool carpet) and trace it back through various hands, following trucks laden with other goods, crossing over border checkpoints with nary a glance by the customs guards, in order to seek out its production site in Tibet, where I would eventually find the nomads that herded the sheep that shed the wool.

This, of course, turned out to be both embarrassingly romantic and methodologically impractical. As I began to inquire about the origins of commodities in the marketplaces of Kalimpong and Lhasa, I discovered that each had such a dense web of far-flung networks of people involved in its production, transport, exchange, and consumption that it was nearly impossible to even begin such a project, partly because of limits in time and funding. A carpet, for instance, could be made with chemical dyes from Switzerland and India and wool from three different parts of northern Tibet and New Zealand, woven by women in a town outside of Shigatse, and made with a design commissioned by a wealthy businessman in Beijing or London. Although tracing the origins of the carpet turned out to be unfeasible because of the complexity of the far-reaching geographical linkages, what I actually found more limiting were the restrictions and the obstacles—for instance, passport laws, roads that were washed out, and permits needed—that prevented nearly anybody (nomads or anthropologists) from actually being able to "follow a thing" from beginning to end.

By switching my methodology from following the paths of objects to staying put in Lhasa and Kalimpong (and later Kathmandu), I was able to interview traders, many of whom departed and returned from trading that very day. I realized that our own immediate, lived experiences of moving and fixing in a volatile social and economic environment could bring together the people, the places, and the things of this study. Assembling a story of traders who cross the national boundaries of China, Nepal, and India necessarily involves crossing disciplinary boundaries. Interdisciplinary themes such as commodity chains, borderlands, and globalization bring together various strands that are crucial to the disciplines of anthropology and geography, often thought of as having specific and separate theoretical canons.

But such an approach loses the kind of ethnographic specificity or depth particular to a long-term residency in a single village of a few hundred people. Studying moving people in multiple towns is not without its many difficulties; like my original intention to follow the paths of objects, inquiring about the details of every single trading link was impractical (for instance, I never found out where Nepali traders went to purchase wholesale goods in Bangkok, or much about the kinds of vegetables sold by the women who routinely ducked under the Nepal-Tibet border barrier). Nevertheless, the peripatetic nature of this book reflects the real frustrations of how contemporary cross-border movement rubs up against hegemonic state apparatuses. To take a rather self-reflexive (or perhaps self-centered) example, because of bureaucratic issues between the United States and the Indian departments of education, I waited on my Indian research visa for nearly a year. This meant that I ended up diverting six months of research time to Kathmandu, which was not included as a research site in the original proposal. Furthermore, due to strikes, curfews, and increasing violence during the peoples' movement against King Gyanendra in Kathmandu in late April 2006, the US Embassy enforced an "ordered departure" call, informing all US citizens registered with the embassy to leave Nepal as soon as possible (despite the fact that Gyanendra had already conceded and there was a low possibility of further violence). We were told that there were only three cities where we could wait out the call for the ordered departure: Bangkok, Dhaka, and Delhi. Dhaka and Delhi required visas ahead of time, and at that point, we were told that the embassies were crammed with foreigners still trying to leave Nepal. I moved again, to Bangkok. It was weeks before the US Embassy retracted the ordered departure and we were allowed to return to Nepal. I use this personal account to emphasize the fact that tensions between movement and halting in the face of state-led restrictions are part and parcel of the intensity of contemporary global connections, and are explicitly geographical. For those who need to cross borders more regularly, these disjointed experiences of time and space, of interminable waiting and diverted movement, also come together to give the impression that the state is "thinglike," coherent, and powerful in its control over space and the movement of individuals (Ferme 2004). Everyday encounters with the state are characterized by the frustrating feeling of being caught in a "warp of space and time," where the "abstract guilt as an individual before the law" is often a result of severe inequality and differentiation, set against the experiences of those who get to cross with less scrutiny (Secor 2007: 49). However, the very nebulousness of being in an ambiguous "warp of space and time" is also important for potential possibility and the creation of diversions around border regulations, as chapter 4 shows.

My expository approach thus mimics the peripatetic, moving-and-stopping-and-moving-again nature of my research, and I return to this parallel again in the final chapter. It does not follow a strict chronology, nor is it divided into chapters representing separate geographical locations. Rather, I am indebted to writing such as that in Anna Tsing's *Friction*, emphasizing the collaborations, tensions, and strange bedfellows that emerge from global encounters, or in Tsing's words, "zones of cultural friction that rise out of encounters and interactions" (Tsing 2005: xi). This combination of people, places, and things working in contradictory movements is what leads to the creation—and diversions—of places, routes, and regions.

More than a decade after Marcus's review article, the conditions for doing fieldwork have changed significantly. In fact, Marcus has more recently criticized the "too-literal understanding of multi-sitedness as simply following objective processes out there by some strategy," likening the initial process of multi-sited ethnography more to an awkward Rube Goldberg machine where the design is developed by the ethnographer and her engagements with "epistemic partners" (Rabinow et al. 2008: 70). Such methods "emphasize the emergent quality of ethnographic knowledge, almost in the sense of a logic of discovery, a discovery of connections, and this practice can only evolve circumstantially, namely in the nitty-gritty arena of fieldwork, in encounters with local others" (Rabinow et al. 2008: 70). Accordingly, this introduction serves as a theoretical background to the emergent ethnographic accounts and analyses that follow. In the next chapter, I provide several "snapshots" of the multiple paths and diversions in the social and economic history of the Lhasa–Kalimpong trade route, complemented by ethnographic accounts from older traders who were operating prior to the closing of Nathu-la in 1962.

CHAPTER ONE

Middlemen, Marketplaces, and Maps

"This is the only Tibetan News Paper published in India, it is read by all the high Lamas, Officials, and leading traders in Tibet, Sikkim, Bhutan, Darjeeling, Northeast Assam, Kashmir Ladakh, Almora, Kulu Himachal Pradesh, Gharwal & Nepal."

From a brochure listing advertising prices in the *Tibet Mirror* for September 1954

Trade and the *Tibet Mirror*

The *Tibet Mirror* (*yul phyogs so so'i gsar 'gyur me long*), a twentieth-century Tibetan-language newspaper published in Kalimpong, had commodity listings in nearly every issue of the newspaper from its start in 1925 until its demise in 1963. These listings gave its readers an idea of what the "market prices in Gold and Silver from Calcutta" looked like, as well as the prices of common items brought from Tibet to Kalimpong. The newspaper clipping below is from the November 24, 1956, edition of the *Mirror* and lists prices for various commodities traded between Lhasa and Kalimpong (figure 5). The items were priced in rupees per *mon do* (maund, a bulk weight of approximately 40–80 kg) and included the following prices per maund: white yaks' tails (800–900 Rs, about US$167–188), black yak tails (440–480 Rs, $92–100), Tibetan pig-hair bristles for hair brushes (20 Rs, $4), musk without skin (60–70 Rs, $13–15), best-quality snow leopard fur (60 Rs, $13), best-quality fox fur (10 Rs, $2), and best-quality marmot fur (1–4 Rs, $0.20–0.84). Mule caravan transport costs were also listed: the journey from Phari to Kalimpong cost 18–19 Rs ($4); Kalimpong to Phari 80–90 Rs ($17–19).

The *Tibet Mirror* was established in Kalimpong in 1925 by Gergan Dorje Tharchin.[1] Tharchin was a Christian missionary from Kinnaur, in eastern Himachal Pradesh, popularly named Tharchin Babu (*babu* is a term of respect

TABLE 1. Market Prices in Gold and Silver from Kolkata, November 24, 1956.

Best gold per tola*	106-8-0
Silver per 100 tolas	176-4-0
Gold coin, each	69-12-0
Best-quality wool from Lhasa per maund	170–
White yak tails per maund	800–900
Black yak tails per maund	440/480/
Tibetan pig-hair bristles, each	20
Musk without skin, per tola	60–70
Musk with skin, per tola	34–35
Best-quality snow leopard hide	60
Best-quality lynx hide	40/80
Best-quality red panda hide	80
Best-quality fox hide	10
Best-quality marmot hide	1–4
Transport costs from Phari to Kalimpong	18–19
From Kalimpong to Phari	80–90
From Gangtok to Phari	70/80

*A tola is an Indian unit of weight roughly equivalent to 11.66 grams.

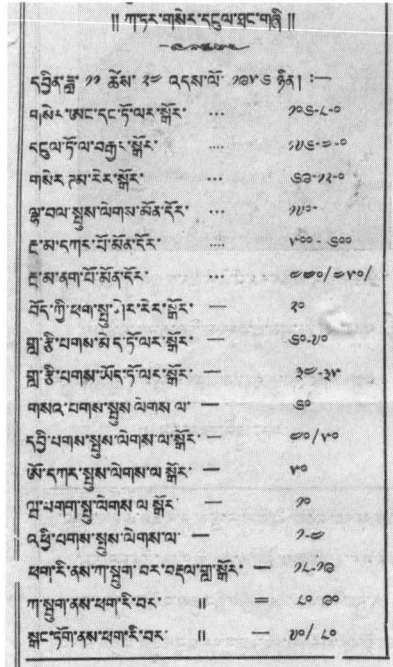

FIGURE 5. List of commodity prices in the *Tibet Mirror*, November 24, 1956. Tharchin Collection, Columbia University Libraries.

for men in several South Asian languages). His newspaper had a small but far-reaching distribution, selling approximately 200–500 copies from Amdo (northern Tibet) to Assam (northeastern India) until 1963. It was transported by the same mule caravans that carried the yak tails and musk listed in the clipping above and was disseminated to medium- and large-scale merchants, traders, and intellectuals living in the towns dotted along the trade routes (Fader 2002: 282). The newspaper provided world and regional news, children's stories, a science section, proverbs, and maps; specific examples include advertisements for wool carders and for "Asiatic Soap," articles about trade disputes, articles about bad weather damaging the wool in the caravans, and notices announcing changes in state trading policies (see figure 6[2]).

These columns about trade in the region, as well as regular listings of trade prices, demonstrate that such "transnational" trading practices have long permeated the economic life of the towns along the Lhasa–Kalimpong trade route. Trading news and various types of information—political, economic, and cul-

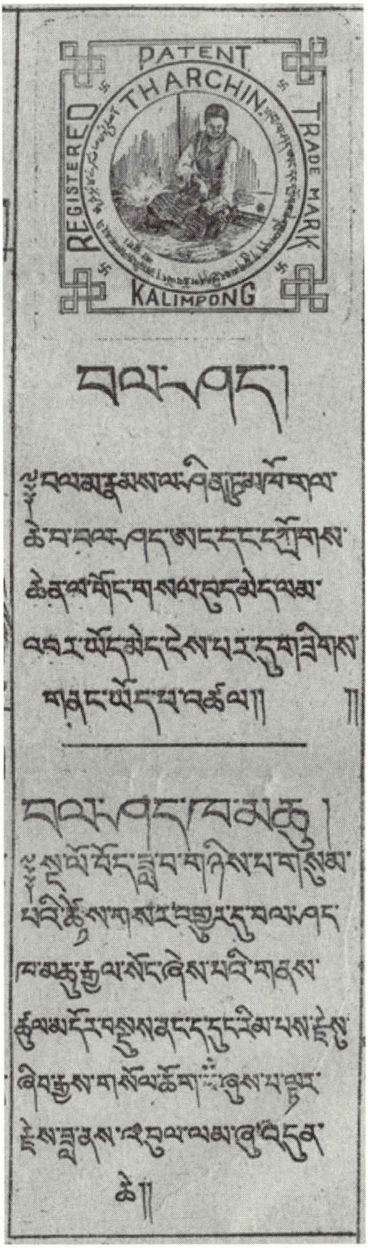

FIGURE 6 An advertisement for Tharchin's brand of *bal shad* (wool carding combs), which were manufactured in Kalimpong (note the Urdu, Hindi, and Tibetan on the stamp). Tharchin Collection, Columbia University Libraries.

tural—traveled along the route with the wool itself; Tibet at this time was certainly not remote in the way that Europeans may have believed, and was in fact very much part of a regional and international economy. To take one example, yak tails—particularly white ones—were (and are) of quite high value because of their use as Hindu, Jain, and Buddhist ritual implements, as flywhisks, and as wigs and Santa Claus beards in Europe and the United States. As a result of North American trade embargos on China in the 1950s, Tibetan exports such as yak tails began to gather slightly more international demand around this time. According to an article titled "Shortage of Yak Tails Worries Santa" in the University of British Columbia student newspaper in 1958, "One Mrs. Cox, who runs a costume supply shop in town, states she has only one Santa Claus suit in stock and that she has had to raise the rental price from $6.50 to $7.50" (Ubyssey 1958: 5). Today, yak tails remain a valuable commodity; costumes for the Orcs, characters in the *Lord of the Rings* films, were made of yak hair, and Santa Claus beards made of white yak tails fetch upwards of US$300.

A deeper understanding of the social and geopolitical history of trade between Tibet, India, and Nepal—as well as trade connections to such larger powers as Britain and the United States—is integral to how we understand the region's present economic shifts. While geopolitical strategies have affected the direction of the trade routes—for instance, the eighteenth- and nineteenth-century British push to "open" Tibet as a market for goods and development and the Chinese "Develop the West" campaign (Cn. *Xibu da kaifa*) of recent history—traders themselves will often have qualitatively different experiences of such changes.[3] Plans to "open" a region, route, or border ignore the fact that the territory has almost never been completely "closed" to actual inhabitants living and trading in the area in the first place. What's more, this choice of words—to "open up" or to "close off" an entire region or route—should not be ignored geopolitically. States or hegemons are seen here as the ones in control of access to the land, not the individuals or communities who live in or move through these areas.

This chapter lays out the historical background to the political, social, and economic climate along the trade route between Lhasa, Kalimpong, and Kathmandu and is supplemented by ethnographic accounts from older traders who were operating prior to and just after the closing of Nathu-la in 1962. I focus on snapshots of specific moments that are significant to the rest of the book, moments that raise questions about competing spatial representations, social mobility, and the strategic directionality of trade. These snapshots take place during roughly three periods. The first period coincides with the British strategy to map the Himalayas as an adjunct of India and to "open" the Chumbi Valley in

Tibet for trade between the late eighteenth and early twentieth centuries. The second period covers the peak and decline of the Tibetan wool trade, the resulting stratification of traders, and the introduction of new kinds of transport along the trade route in the mid-twentieth century. The third period follows the rerouting of the Tibet–India trade through Nepal after the 1962 Sino-Indian border war, culminating in a ghost story that appears to link border disputes from the past with the more contemporary drive to increase Sino-Indian border trade.

The intertwining of economic and geopolitical strategies is a major part of the larger historical picture of macro world histories. Trade cannot be studied as a phenomenon in isolation from state policies, nor can the practices and lives of traders, which often reveal a very different trajectory from those policies. A trade route is never simply "there," displaying a simple, recognizable shape over the course of centuries; groups of people may follow the same paths, but there is always more than one route or conception of the route being forged at a time. The Himalayan landscape has always been molded, reformed, argued over, rerouted, and produced by people. This is a dynamic story of fixity and movement, of trade restrictions established by state powers and how various competing parties respond by maneuvering around them. I hope to show, through historical and contemporary accounts, "the dynamic connections between political powers and geographical knowledges of different sorts" (Harvey 2001: 233).

Mapping the Himalayas: British Designs on Tibet

Tibet has often been portrayed by Western travelers and scholars as remote and otherworldly. Even in the March 2008 media reports on anti-Chinese unrest in Tibet, the region was depicted as only very recently opening up to the outside world. Take for instance headlines such as "In Remote China, Tibetans Break Silence" and "Why Is Remote Tibet of Strategic Significance?" (Anna 2008; Reuters 2008). Tibet has, however, been part of a vibrant network of inter-Asian caravan routes ever since the seventh century, when Indian scholars introduced Buddhist texts to Central Asia and China, and pilgrimage routes paralleled the tributaries of what historians now consider the northern Silk Road (Klimburg 1982: 33). Tibetan musk was even purportedly traded between India and the Roman Empire (Beckwith 1977). Rather than a single "road," these numerous trading and information networks ramified in several directions between the seventh and tenth centuries: textiles and gems came from northern China, tea arrived in the form of bricks from Yunnan, raisins came from Khotan, seashells from south India, and pashmina shawl wool from Ladakh.[4]

Although trade has risen and fallen throughout inner Asia over the centuries, it peaked significantly in the early to mid-seventeenth century, when trans-Himalayan trade coincided with several geopolitical factors—in particular, the rise of Mogul India and the establishment of Western maritime powers such as the Dutch East India company (Boulnois 2003: 135, 137). In 1642, for instance, the fifth Dalai Lama "presided over a large Tibetan state from Kham to Ladakh and the development of Lhasa as the administrative, religious, and commercial capital," while the rise of the Moguls made the Malla kingdoms of Nepal prosperous; they acted as middlemen for trade between India and Tibet (Boulnois 2003: 138).

The British East India Company became interested in Tibet in the late eighteenth century, around the time the monopoly started to show signs of slowing down and subsiding. As the company started to lose its grip on trade in Asia, it was forced to think about a locational shift; it thus began to specialize in overland textile trade through the "blank" areas of inner Asia (Arrighi 1994: 248). When the Gurkha conquest of Nepal shut off trade between Tibet and India in 1769, the company made an explicit move to introduce British commodities such as Manchester textiles and Indian indigo to the potential new market of China via the "back door" of Tibet (Cammann 1970: 144–45).

The task, then, was for the British to explore the "the nature of the road between the borders of Bengal and Lhasa" to prolong their economic and territorial expansion (Bishop 1989: 82). In 1774, Warren Hastings (governor-general of British India from 1773 to 1785) asked George Bogle, the first British envoy to Tibet and Bhutan, to send a team to investigate the area. Bogle was given a list of items to procure, including pashmina goats, two yaks, walnuts, rhubarb, ginseng, and any animals that he happened to find "remarkably curious" (Bishop 1989: 30). According to Peter Bishop's account of the history of Western spatial representations of Tibet, while Tibet had a general location (somewhere between China, India, and Central Asia) in the British imagination at that time, it had very little shape (Bishop 1989: 30). Bogle's initial meetings in Tibet were short-lived, partly due to his untimely death by drowning, but British interest in "opening" Tibet for commercial purposes was taken up again a century later. Echoing earlier trade reports from the East India Company, Francis Younghusband, the leader of the 1904 British invasion of Tibet, recalled that in 1873, "besides tea, the Bengal Government thought that Manchester and Birmingham goods and Indian indigo would find a market in Tibet, and that we should receive in return much wool, sheep, cattle, walnuts, Tibetan cloths, and other commodities" (Younghusband 1910).

Hand in hand with the need to gain scientific knowledge about the flora and fauna of the "unknown" region beyond the existing British empire was the need to create a uniform, structured survey of the territory: to "know" India, minus its diverse specificities, of course (Edney 1997: 324). At the same time that we are reminded about the power relations involved in the actual practice of state- or government-led exploration and the subsequent creation of a map of a region or a route, there seems to be a certain state-based conjuring taking place, entangled with the magic and mystery of exploring the unknown. "What seems crucial here is what this tells us about the state; how it needs this theater with its magic, but needs it disguised as science, and how important the domination of nature is to such theater" (Taussig 2004: 198). In other words, the exploration and cartographic representation of Tibet often tells us more about the establishment of the shape of the British empire than it does about the physical landscape of the region.

Although the nineteenth-century exploration and geographical mapping of the Himalayas through the British Great Trigonometric Survey is already widely documented, I refer to it here briefly, as cartographic knowledge is often an elite and uneven form of power.[5] The Himalayas, as a major part of the survey, began to be seen as connected to India and therefore came to be viewed by the British as a "rational" and unquestionable part of the British geography (Bishop 1989: 89). Attempting to attach Tibet to the map of British India, motivated in part by fears of Russian expansion during the period known as the "Great Game," the British colonial authorities harnessed spatial knowledge from inhabitants in the areas where surveys were undertaken, somewhat ironically mapping the "new world by erasing local histories" that were dependent on "local knowledges" (Pickles 2003: 119).

In 1863, the Royal Geographical Society clandestinely trained and sent out Indian and Sikkimese surveyors into Tibet (central Tibet followed an isolationist policy in the nineteenth century, not allowing foreigners into its territory). The surveyors were disguised as traders and given special mapping instruments that had to be concealed; surveyors used Buddhist prayer beads, for instance, while walking along, pretending to murmur mantras for each bead counted. Although they normally have 108 beads, an auspicious Buddhist number, the British instruments were made with 100 beads so that they could measure paces evenly. Buddhist prayer wheels, instead of having scriptures inside, would have slips of paper for recording compass bearings. Compasses and field books were to "look as 'un-English as possible,'" and mercury was inserted into coconuts for clandestinely measuring altitude (Madan 2004: 95). According to members of the Royal Geographical Society, the pundits' "labours, if successful, will make . . . our present stock of Geographical and general knowledge of the vast unexplored countries which lie between India, China and Russia in Asia, most important" (Madan 2004: 143).

The establishment of British Trade Marts in Tibet provides another example of the close relationship between early twentieth-century British geopolitical strategy and British economic advancement. Soon after the survey was completed, an official note stating that the Russians might also want to trade with Tibet was the immediate excuse for the 1904 invasion of Tibet led by Francis Younghusband (Cammann 1970: 147). The invasion was accompanied by the rhetoric of "opening up" the Chumbi Valley (the valley between present-day Sikkim and Tibet) for trade, as if there were nothing there before. This movement toward the "opening up" of the borders between Sikkim and Tibet produced a situation that was actually contrary to this phrase's connotation of unimpeded, unrestricted mobility. It went hand in hand with the signing of a number of trade treaties put in place in order to establish British Trade Marts in the Tibetan towns of Gyantse, Gartok, and Yatung, as well as the solid demarcation of boundaries between Sikkim and Tibet.[6] However, local inhabitants offered some resistance to the trade marts and boundary lines. According to Younghusband's account:

> The "Chief Steward," the sole Commissioner on the part of the Tibetan Government for reporting on the frontier matter, "made the important statement that the Tibetans did not consider themselves bound by the Convention with China, as they were not a party to it." He reported further, that the Tibetans had prevented the formation of a mart by building a wall across the valley on the farther side of Yatung, by efficiently guarding this and by prohibiting their traders from passing through. Mr. Korb, a wool merchant from Bengal, had come to Yatung to purchase wool from some of his correspondents on the Tibetan side, who had invited him thither; but the Tibetans prevented his correspondents from coming to do business with him. (Younghusband 1910)

In a 1930 guidebook for British tourists to Tibet, David MacDonald discusses a similar incident. "When the milestones were first erected on the Phari–Gyantse road, the local people invariably destroyed them, alleging they were gods put up by the British to destroy their faith. Only after many years did this belief die out, and even at the present time some of these cairns can be seen scattered over the plain" (MacDonald 1999: 102).

Although the milestone-markers-as-gods allegation is difficult to verify, this incident clearly points to local attitudes toward the construction of the British version of the trade route. British authorities complained that the "imposing of free trade [was] rather difficult because Tibetans *were used to their own form of trade rules*" (Bell 2000: 256, my emphasis). The trade agencies doubled as administrative units; David MacDonald, a trade officer in Tibet for nearly twenty years, wrote of his duties in Yatung, which "consisted of administering the Trade Mart, caring for the interests of British subjects trading in Tibet, and watching and forwarding reports on the political situation [vis-à-vis China] to the Government of India" (MacDonald 2005: 52–53). Although somewhat cut off from British strategic headquarters in England and India, the British trade agencies (which became Indian trade agencies in 1947 and formally continued with Chinese control over communication operations in 1954) operated in Tibet until the Sino-Indian border war in 1962 (McKay 1997). All of these strategic actions—the establishment of trade agencies, boundaries, trade treaties, and new roads—were intended to facilitate unimpeded "free trade" for the British while restricting the movement of existing traders in the region. Despite these restrictions, however, local traders continued to go about their daily activities, even if it meant destroying border markers or redirecting their trade journeys around such limitations. Thus, when the border markers were established so that the Chumbi Valley could be declared "opened" by the British, "the inhabitants were so angry they kept destroying the border markers, and trade happened anyway despite the restrictions" (Bell 2000: 61).

The Peak and Decline of the Wool Trade

In a teashop in Lhasa, I had a conversation with a woman in her eighties who used to work as a merchant in a store on the Barkor (the main pilgrimage—and shopping—circuit in Lhasa) while her father and other male family members led mule caravans between Lhasa and Kalimpong. The woman said she was from a "small" trading family. People in Lhasa, Kalimpong, and Kathmandu often referred to themselves and others in Tibetan, English, or Nepali as "small" or "big" traders (or more rarely, "medium-sized"). Generally, the small traders are those who maintain a living by exchanging goods within a relatively local radius, and large traders are those who tap into more distant or international networks (but who are not necessarily more mobile).

The woman from the small trading family insisted rather seriously that "selling wool is very dangerous. Because if you have good luck, you will be very rich, but if not, you will be broke." The Tibetan term she used for "broke" was *rkub rdib*, which could translate literally into English as "your ass has collapsed." Traders in the 1940s were subject to huge risks in the transport of wool. Not only was the quality of the wool dependent on weather conditions (heavy rains along the trade route could ruin a whole caravan-load of wool), but trading caravans were also common targets for thieves, as they would often be carrying items of value. The elderly woman in the teashop remembered one story about the risks involved: "In the countryside, on the way to Chushul, some of the *bdal ring* [muleteers who took long journeys] would have goods stolen. One iron trunk had sixteen boxes of candies in it. The thieves put stones in the boxes instead. They stole cotton, too, and because the outside of the cotton was tied in a bundle, they pulled out the cotton goods in the middle. If you aren't careful with calculating everything in your possession . . . when you get to Lhasa, you must not lose anything or get anything stolen."

Although the majority of traders often had much to risk, a few Tibetan families were able to establish far-reaching, successful trade networks in the early twentieth century. Fernand Braudel, in his account of capitalism and economy in Europe, showed how increases in the circulation of capital led to increased divisions among traders, producing a hierarchy of small and large traders. As trading elites grew more and more prosperous, they became an aristocracy (Braudel 1982: 68). In Lhasa, the three main trading families—Pangdatsang, Reting, and Sadhutsang—constituted such an aristocracy, and in fact, the families' names were commonly abbreviated and mentioned jointly as "Re Pa Sam"

FIGURE 7. Mule caravan loaded with Tibetan wool, near Phari (?), circa 1930s. Photo by K. C. Pyne, Kalimpong Stores (Kodak).

FIGURE 8. Traders on the main road of Kalimpong, circa 1930s. Photo by K. C. Pyne, Kalimpong Stores (Kodak).

(*re spom ga gsum*; McGranahan 2002: 104).[7] All were originally from Kham, in eastern Tibet (current-day Sichuan, Yunnan, and Qinghai Provinces), and grew to obtain exclusive contracts with wool purchasers in the United States and Europe. Charles Bell, political officer of Sikkim, wrote that "during my time in Tibet [1920–1921] the chief Tibetan merchant was a man named pom-da-tsang [Pangdatsang]. He had branches in Calcutta, Shanghai, Peking, and Japan. He exported mainly wool—but yak tails, too" (Bell 1928: 130). Similarly, one member of a medium-scale trading family who lived in Yatung spoke about those three main trading families in the 1950s: "These families had trade stations in Changtang [the nomadic shepherding area in Northern Tibet], like district offices. Pangdatsang members were so wealthy they even had their own private banks in Kalimpong, and they even had a factory that separated the wool into different qualities, into white and black, which were then packaged for export." While talking to some older traders, I heard—but could not later verify—that a major American car manufacturer had signed contracts with the Pangdatsangs to use Tibetan wool for their automobile mats.

Until Indian independence in 1947, the British-appointed trade agents in Tibet kept diaries, summaries of which were reported annually to the British Indian government. Most entries are rather dull, marking the visits of British and Tibetan officials, noting miscreants who cut down telegraph wires, listing prices of wool and tobacco, but they also provide telling indications of the economic divisions between the monopolizing trading families and their less wealthy counterparts. In 1944, Sonam Tobden, the Tibetan agent stationed in Yatung, wrote the following: "The three prominent Tibetan traders have brought sufficient wool in Tibet and are now transporting them [loads of wool] to Kalimpong. Most of the petty traders had to sell their wool at low price at Phari owing to the scarcity of transport animals. They suffered heavy losses in the wool trade as they have paid higher price for wool in inner Tibet." And in 1946, Tobden recorded: "The big Tibetan traders who have godowns [warehouses] at their disposal at Phari and Kalimpong for the storage of the wool managed to earn much more money than the petty Tibetan traders who have no such godowns. Added to this, the petty traders who usually resort to borrowing large portion[s] of their capital at usurious rates of interest were unable to hold their wool for a considerable length of period even when there was good prospect of substantial increase of the price of wool in the near future."

The local petty traders and merchants were clearly frustrated with corrupt middlemen and elite traders taking advantage of them, as also indicated by an anonymous letter addressed to the Secretary of the Foreign Affairs Department in New Delhi in 1946. The author writes,

> Tibetan merchants who get permits of goods at Kalimpong resell them at double or trible [sic] prices to people who take these goods down to India and sell them at very high prices to Indian people. Thus huge black marketing is being done and any appeal to the local authorities is proving fuitile [sic] because the local officials at Kalimpong and other places are cooperating with the black marketeers and in spite of specific instances being given no drastic action has been taken. The public suffers and these highly-placed plotters are becoming enormously rich.

It doesn't seem that the British paid much heed to this letter, as a handwritten note is attached to the document, stating, "Were all these allegations true, which I very much doubt, there is nothing we can do with an anonymous letter." But if the elite trading families had the monopoly on the wool trade in the 1940s, their position was made a bit more precarious by the revolution in China. By the time the Communist Chinese troops entered eastern Tibet in 1950, the wool trade was already badly affected by the halt in sales to the United States, as the United States began to cut off all economic ties with China and Chinese territories. According to a 1951 *New York Times* article, "It is learned from trading circles that an estimated 1,600,000 pounds of wool have been stockpiled here [in Kalimpong] against the season's normal anticipated import of 8,000,000 pounds. Out of this, nearly 2,000,000 pounds will not arrive from Kham in Eastern Tibet, which is now under control of Chinese Communists" (*New York Times* 1951a).

The Tibetan economy being tightly integrated with the Indian economy at this time, the Chinese occupation led to rapid geo-economic transformations on different spatial scales (Shakya 1999).[8] As the United States cut off economic ties with China, the big traders who held monopolies on wool quickly lost most of their profits, and four million pounds of wool lay rotting in Kalimpong warehouses. At the same time, however, smaller-scale traders suddenly experienced a boom in business. The beginnings of Chinese rule in the early 1950s led to a thriving business for many traders, not least because certain products from India (for example, watches, surgical instruments, and nail clippers) were much desired by the People's Liberation Army (PLA) soldiers stationed in Lhasa (Radhu 1997). In the 1950s in Lhasa, Chinese soldiers were said to be "avid shoppers," buying watches and fountain pens and bringing in Chinese silver coins (*da yuan*) by the truckload (Tuladhar 2004: 92–93). There was in fact a huge illicit trade in Chinese Kuomintang silver coins brought to India at the time. As the coins were phased out in China, they were taken to India to be sold and melted down, since the silver content in the coins was worth more than the coins themselves. Mule caravans from Tibet were said to arrive in India nearly empty (except sometimes for the coins) and be loaded up to return to Tibet with Indian textiles, batteries, cigarettes, and sweets, leading businessmen to state that "if the present slump in Indo-Tibetan trade continues, Kalimpong can be written off as the most important trade center" (*New York Times* 1951b).

The Chinese were quickly able to redirect much of the Tibetan economy away from the south—away from India and Nepal—and up toward Beijing; the State Bank of China in Tibet, which offered interest-free loans to traders and businesspeople, was established in 1952, and PLA-built roads were completed in 1953 (Shakya 1999: 135). The US market, which usually received around 70 percent of the annual eight million pounds of export wool from Tibet, declared that it would not accept any shipments after February 29, 1952. Russians and Czechs then began to buy Tibetan wool, and China stepped in to purchase all eight million pounds of surplus wool "to help the Tibetan government tide over the current financial crisis caused by a slump in wool trade" (*New York Times* 1952). TN Sherpa, of the Tibetan Traders Association, took over all the individually owned warehouses.

As the wool supply that was formerly brought daily from the nomadic areas of central and northern Tibet to India was nearly cut off, the direction of the trade reversed quite dramatically for a few years. What had primarily been a north-to-south trajectory of trade became a south-to-north one. Wangchuk's story exemplifies the eve of this shift. A man in his sixties who lived in the trading town of Yatung, Wangchuk brought *rgya zog* (sugar, textiles, and food—literally "Indian goods" in the Tibetan language) to Tibet and wool to India in the 1950s:

> You could buy many things from foreign countries—mostly Europe—in India and bring them up to Tibet. By 1955 there were roads in Dromo [Yatung/*Gro mo*], and from 1951, the Chinese army really needed supplies. Much of these came through Dromo, and from that year on, business became bigger and the town became wealthier. When I went to Beijing in 1956, I thought it wasn't developed at all, and I was surprised at how poor Beijing was compared to Dromo. The three big [trading] families had a federation, an association for businessmen [Cn. *she hui*]. Before 1951, you didn't need a permit to travel back and forth between Tibet and India, but from 1951 onward you needed a permit; still, it was very easy to travel. It took thirteen days from Shigatse to Kalimpong, and it took seventeen days from Lhasa to Kalimpong.[9]

And yet, toward the end of the 1950s—after the crushing of a Tibetan rebellion and after the Dalai Lama's decision to leave Tibet in 1959—traders in India began to feel some of the adverse effects of the Chinese-directed new economy. Mr. Agarwal, in his mid-eighties, is said to be one of the oldest Marwaris still living in Kalimpong. He used to be a member of the Kalimpong Chamber of Commerce, selling ready-made clothes and textiles from Bombay, Surat, and Ahmedabad. He said: "In 1959 things started to become very difficult, business was very bad. I remember one story of a merchant in Kalimpong holding 100 *do po* [loads] of chubas [traditional Tibetan garments] for someone in Tibet who was supposed to come and barter wool for them, but this happened right around 1959, when things were bad. He waited and waited and the guy never came; he really lost then, he wasn't able to make a profit at all because few people in Kalimpong wanted these things."

At least six traders in India and Tibet have reported that around 1959, "trade virtually stopped." Those who could do so migrated to other places, such as Kathmandu, where business was more lucrative. One reason for the decline in trade was that toward the end of the 1950s, PLA-led road-building projects from major Chinese cities into Tibet were completed, and supplies no longer needed to come from India. One former trader now living in Kalimpong lamented, "In 1959 everything stopped. In the mid-1950s, the Chinese built roads, and when the roads were made, business went down." In addition, after 1959, Tibetans in

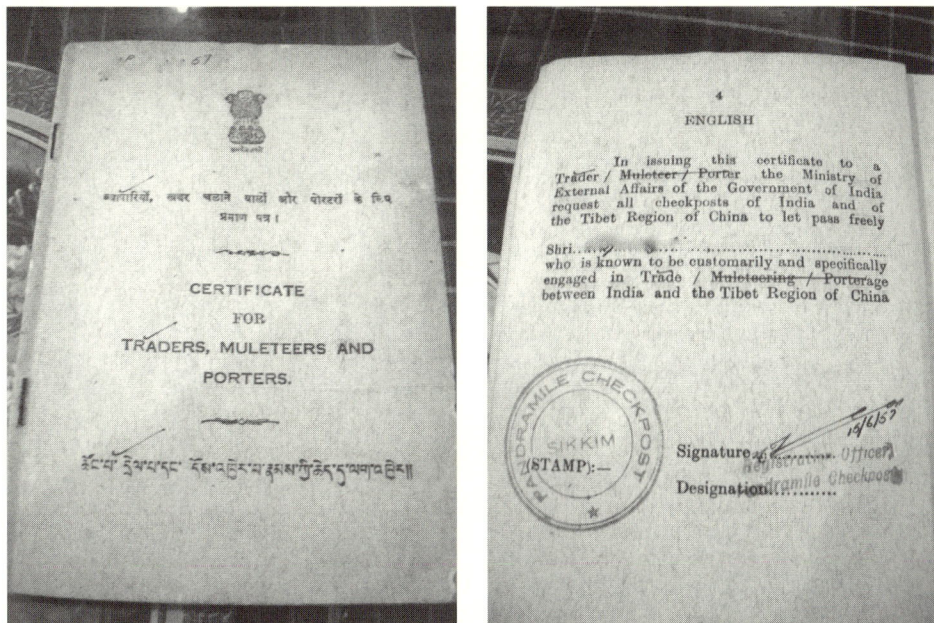

FIGURE 9. A trading permit from 1957. Note that the three languages in use at the time were Hindi, English, and Tibetan. Tina Harris.

Tibet who wished to trade between Tibet and India had to obtain three separate difficult-to-obtain permits: one from the Government of India, one from Sikkim, and one from Kalimpong. He sighed, "In a way, it wasn't worth it."

Changes in Transport: Roads and Motorized Vehicles, 1950s–1960s

A trade route is often conceived of as a single line on a map where commodities are transported from point A to point B. Of course, the route between Lhasa and Kalimpong has never been direct. It has been cut off, elongated, and diverted, and people do not necessarily follow the same path as the products that travel along the route. For instance, when asked about the trade route, Tibetan traders would not speak of it as a simple point-A-to-point-B route, but would instead recite the names of towns along the way. One elderly trader said the following: "The trade route from Dromo to Lhasa was Dromo, Phari, Dhuena, Khama, Nyero, Gyamo Naga (Ralung), Nagatse, Draknalang, Chu Shue Gang, and Lhasa on horse, mule, and yaks. From Dromo to Kalimpong, it was Dromo,

Nanga, Zelebla, Kubub, Gnathang, Lingtam, Rhenock, Pedong, and to Kalimpong on mules and horses. I liked the fact that my business fetched me good money, but the tiring trek from Chumbi to Kalimpong was something I detested."

It is difficult to standardize place names along the route because they tend to have numerous pronunciations, transliterations, and spellings in English, Tibetan, and Chinese. In figure 10, I have mapped the town names as the trader wrote them down in English. However, other names or spellings that are commonly used are as follows: Kubub (Eng. Kupup), Zelebla (Eng. Jelep-la), Nanga (Tib. *nags ra?*), Yadong (Eng. Yatung, Tib. *gro mo*), Phari (Tib. *phag ri*, Cn. Pali), Dhuena (Eng. Tuna, Tib. *dud sna*, Cn. Duina), Khama (Eng. Khangmar, Tib. *khang dmar*, Cn. Kangma), Nyero (Eng. Nyeru, Tib. *nye ru*, Cn. Nieru), Gyamo Naga (Eng. Ralung, Tib. *ra lung*, Cn. Relong), Draknalang (Tib. *brag dkar la?*), Chu Shue Gang (Eng. Chushul, Tib. *chu shur*, Cn. Qushui). The question marks reflect the uncertainty of the exact names or spellings, according to traders and friends in the region. In addition, two of these place names—Nanga and Draknalang—are possibly the names of mountain passes, not towns. This kind of representation of the route demonstrates one of the many different ways that traders produce oral histories of trading space, not to mention the historical and political significance of using the Tibetan place-names instead of either the more recent Chinese ones or the names that were given in English during the British colonial era.

Another reason the route was represented by the trader in this way—from small town to small town—is that the mule caravans were leased and loaded in stages. There was more risk of tiredness, injury, or deaths among the animals if they had to travel the entire route; moreover, those who leased out the animals could make a good profit while remaining in the town. In Lhasa, a woman in her late seventies explained how the *bdal gla*, the transport fee paid to those who leased out the mules, worked. She said that you would pay for the mules in stages, for example, from Kalimpong to Phari; but a longer *bdal gla*, say one from Phari to Lhasa, was significantly cheaper. There was a particularly short trip, she said, beginning in a village somewhere around Gyantse and Kangmar—"maybe Tramalung?"—where the people were really shrewd and were able to glean a high profit from such a short *bdal gla*. The competition was high: "There were so many stops along the route, so many groups will take the goods along." She recalled that at each village there had to be a lengthy negotiation in order to settle on a suitable fee, and that one person—a *mi ba ru* [a middleman]—would be responsible for all the negotiations until they were completed. This person would discuss the load and how much it would cost, and then

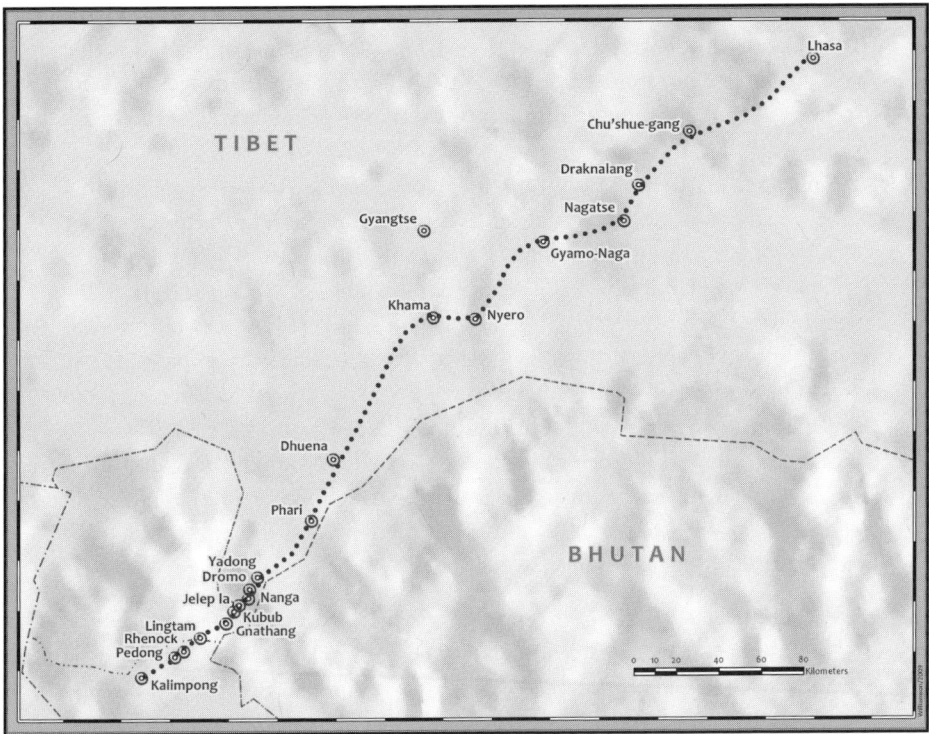

FIGURE 10. The caravan route described by a Tibetan trader, remembered from his journeys in the early 1950s. Doug Williamson, Hunter College Department of Geography.

would get a profit. Each day, she recalled, at least ten or eleven loads were sent out from near her home in Lhasa, mostly with donkeys. She insisted that "if you use mules, they are a little more expensive than donkeys. Also, a longer route for *bdal gla* is cheaper because it doesn't have to go through so many people who get a profit [but you could also lose the animals more easily]. From Phari to Dromo, it would take one day, and from Dromo to Kalimpong with mules, maybe two or three days."

The daily trading activities and experiences of this kind of "friction of terrain," then, are fractured and marked by more of a stopping-and-starting kind of motion—several different modes of transportation are used, different fees are paid depending on the length of the stage or the efficiency of the middleman, and these are all highly dependent on the weather or the health of the transport animals (not to mention the fact that the same animals are not used along the entire route). Moreover, not a single trader remembered the landmarks along the route in exactly the same way, further exemplifying the variety of ways trad-ers "make" trading routes. As I mentioned, in figure 10, some of the names that were given were not immediately identifiable on a map or in any gazetteers of Chinese place-names, and may in fact be the names of landmarks or mountain passes.

One elderly Tibetan Muslim man remembered the kinds of materials that were brought from Kalimpong around the same time period: watches, clothes, "absolutely everything" went from Kalimpong. And from Tibet, he recalled traders bringing gold, silver, and semiprecious stones to sell in the Barkor, the main marketplace. And yet, based on the goods and trading networks he was part of, his geographical trajectory was somewhat different compared to the map of figure 10. His relatives, who were also traders, went to Kolkata. From Kolkata to Siliguri, they would bring the clothes by railway. From Siliguri to Kalimpong, they would bring the clothes in bullock carts, in wagons, or on horses. And what he remembered was that from Kalimpong to Lhasa it would often take two or three months in total, as there was much snow on the road and occasionally an avalanche. He continued:

> Sometimes people died with the horses . . . There was only one way at that time. There was not more than one route, and no two-way lanes. Sometimes horses on the opposite sides of the road would get into fights, sometimes they would com-promise. There were bandits and robbers as well who would come from the Tibet side, but the businessmen were clever at the time. They would set up their tents but not sleep in them, because they knew the bandits would attack the tents first. The route was like this: Kalimpong, Pedong, Rhenock, Rangli, Gnatang, Phatamji, Zuluk, Kupup, then Jelep-la. That is at the border of India and China. After 1947 all the horses went to Tripai, behind Dr. Grahams' Homes [a well-known school established by a Scottish missionary in the early twentieth century in Kalimpong]. There was a ropeway from Durpin which brought goods—wool—to Lohapul, where there was a train link to Siliguri.

Many older Tibetan traders remembered motorized vehicles first appear-ing along the Lhasa–Kalimpong trade route in the mid- to late 1950s. Because crossing the high-altitude Himalayan mountain passes is quite treacherous, and the roads little more than craggy mountainside mule paths in many areas, there was no way for motorized vehicles to be driven over the mountains to the other side. Instead, trucks would transport the traders and their goods only as far as Phari on the Tibetan side, and only up to Kalimpong on the Indian side. As another elderly trader recalled: "From 1955 trucks slowly came into the picture, though only as far as Phari, and then mules would take the load the rest of the way. Much of the Tashilhumpo (Shigatse) loads were taken by *ther ka* [a horse or mule-drawn cart], in caravans of twenty to thirty. In 1957–58 trucks were more common in Phari."

Again, the divisions in wealth between small and large trading groups during this time were felt acutely. A woman in her late seventies said that although there were more roads and cars connecting Lhasa to other towns in Tibet from 1957 onwards, "the bigger families, like the Tsarongs, had cars before then. They would take horses and mules over the passes, but then the cars would take them down to Kalimpong."[10] One favorite recollection of many elderly traders was the unique method of transporting cars from the ports in Kolkata to Siliguri and then over the mountain passes to Lhasa. The cars needed to be totally taken apart, separated into pieces for individual people and mules to carry over the treacherous passes, and then put back together again on the other side.

One of the traders who remembered transporting the vehicles to Lhasa in the 1950s was Mohamed, a Muslim man who worked for the Bhajuratna—also called Syamukapu—trading company, perhaps the best-known Newar shop to still maintain a presence in Lhasa today (the nickname "Syamukapu" is based on the Tibetan for *zhwa mo dkar po*, "white hat," and is said to refer to the white cap that Bhajuratna Kansakar, the founder of the company, used to wear).[11] As Mohamed recalls, he worked for the Bhajuratna company between 1954 and 1958 and was given about 300–400 Indian rupees per month, which was a considerably good salary. At the time, business was booming in the south-to-north direction between Kalimpong and Lhasa, and one of his tasks was to import bicycles and automobiles from Kolkata into Tibet. His account of reassembling vehicle parts includes a comparison of two of the wealthiest Tibetan trading families from mid-twentieth century Tibet with the Tatas and the Birlas, two of the biggest Indian families who had had the monopoly over the steel, textile, cement, and auto-making industries in India since the late 1800s.

> The car would be in different pieces. Until early in 1958, that was the way we were doing it. From Siliguri, they would have it [the car] in small pieces. They would buy a lorry from Siliguri in different pieces. Some pieces were for the mules and some for the humans to transport, up to Phari. From Phari, we had a little bit of transportation up to Shigatse. The roads were not so good, they were dirt roads. We would assemble the cars [on the other side]. They were getting 8000 Indian rupees to assemble the cars. Both Sadhutsang and Pangdatsang, the biggest, rich-

est Tibetans, were doing that [with the cars]. But Bhajuratna, they would take some loans, they would ask Sadhutsang and Pangdatsang for loans. Yes, they are like the Indian Tata-Birlas. They are the Tibetan Tata-Birlas! [laughs].

The stratification of traders is marked by their different kinds of trading geographies; despite all the political changes in the mid-twentieth century, the wealthy Pangdatsangs and Sadhutsangs, who lost a fair amount of their wool trading profits at the beginning of Chinese rule in Tibet, maintained links with suppliers in Kolkata, and then sold—and of course owned—automobiles as they became more popular in Lhasa. And, as might be expected, the advent of air transportation ended up isolating many other traders because it was extremely costly (van Spengen 2000: 193). For example, a former trader who used to sell goods in Kalimpong from 1956 to 1958 bought his supplies in Kolkata. Although he was able to transport his commodities by plane during this time, this option was limited to only the wealthiest traders or those who were well connected with people in the transport industry. This trader would take an airplane from Kolkata to Siliguri and sit in the cargo area with the luggage, which cost 75 rupees per person. "Back then," he said, "there were no customs, no restrictions; you could buy things as you pleased, no one checked anything. After that, everything changed. Kalimpong changed."

Major shifts in political or ideological control in China and India during the mid-twentieth century did not result in immediate, parallel changes in trading practices for all of those involved; these results were often staggered and difficult to pinpoint while they were happening. Instead, factors such as technological changes in transport and infrastructural transformations seemed to have much more of an impact on the direction and practices of trade; as access to new modes of transport was limited to those traders with far-flung business connections and relative wealth, stratification between "big" and "small" traders grew even wider in the late 1950s.

A few years later, the sealing of the border immediately following the 1962 Sino-Indian War turned out to be another momentous event that led to crucial geographical and social shifts in trade. Although I do not discuss the geopolitical history of the war in any detail, what I find especially significant is how the closing of an international border becomes entangled in the contradictory rhetoric of state security and free trade, making visible the complex tensions between fixity and mobility on a national scale. The next section addresses these tensions through a ghost story that ties anxieties associated with the past events of the border closing to plans for the future reopening of the border.

The Border Baba and the Haunting of the State

During the course of my research, much of my time was spent trying to track down elderly traders or their relatives in Tibet, Nepal, or the northern parts of West Bengal. Twice, I was given a piece of paper with a single name and place on it, for instance: "Tenzin, Kalimpong." Although equivalent to a piece of paper that simply said "John, Ramsgate," it would often be followed by a quick "I think he used to own a shoe shop," which, supplemented by the knowledge of who was sending me, was enough to point me in the right direction. Someone, somewhere, always knew who Tenzin was, and even where he was, though often I would be told that he had passed away, or that he had moved to Darjeeling (where I had just been a day earlier), or that he was working in California. During one of these searches in Sikkim, I was trying to locate someone who was involved in transporting musk from Tibet to India in the 1950s, to no avail. I sat in a tea shop, wondering if I should just make the long walk back down the hill to my rented room. Just as I was giving up and packing up my field notes for the day, someone found me instead. A man in his late thirties came in, saw me looking at my notes, introduced himself as Arjun, and mentioned that he himself was interested in borders. In fact, he was meeting an "expert on borders" this evening; would I care to join them for dinner? Wondering which policy maker or academic might be in town for the weekend, I met Arjun and the border expert at a restaurant where the staff dressed in pressed white shirts and black trousers. There was no one else in the restaurant. It turned out that the "border expert" was a man who had had some experience stationed on several China-India border points (one of them was "a horrible, horrible place," he recalled as he shuddered), as well as on Jelep-la. He then asked if I had heard about the Border Baba. "No," I said. "Tell me more."

Geopolitical tensions on the boundaries of India and China built up steadily between the late 1950s and early 1960s. Following the Chinese government's repression of a Tibetan uprising and the exile of the Dalai Lama in 1959, the Sino-Indian border war in 1962 cut off nearly all trade between India and Tibet. With the tightening of the military presence, India withdrew its trade agencies in Yatung, Gartok, and Gyantse, while China withdrew its trade representatives from Kalimpong and Kolkata. Trading between China and India effectively stopped, and the well-worn Himalayan passes of Jelep-la and Nathu-la fell into disuse (Maxwell 1970: 235). At this point, many Indian traders who had shops in Tibet were given twenty-four hours' notice to leave; one current Marwari trader remembers this sudden economic upheaval when he was a child in Sikkim and his father had a shop in Yatung. He began by discussing how the mule trains left MG Marg (one of the main roads in Gangtok) at 3:00 a.m. for Yatung with four 40 kg packets of grains on each mule, but when the pass closed on June 2, 1962, "whatever they had, they had to bring back in bulk to Gangtok." Other traders recalled similar stories, emphasizing the disruption of daily life. One man remarked, "I found out about it [the pass closing] on 1 June and was able to leave, but other traders who stayed past that date were forced to leave within twenty-four hours." Another former trader currently living in Kalimpong said, "From 1962, business completely stopped, and the border was sealed on both sides. We moved here, to Kalimpong, in 1962. At that time, businesspeople spread everywhere."

Since there seemed to be no imminent future for trade with China through the Sikkim routes, many trading families picked up their belongings and headed to Kathmandu, where they had relatives or business partners. Trade between Tibet and India slowly began to build up again, but now along the diverted route through Nepal toward Kodari, on the Tibet border. In fact, the Nepali king Mahendra had visited Peking as early as 1961 and received a blueprint for a road linking Nepal and China. Committed to creating a route fit for trucks, Nepal employed twelve thousand Nepali farmers (and fifty Chinese engineers) to build the Kodari road and the "Friendship Bridge," enabling a stream of official cross-border trade by 1964.

> Replying to Indian charges that the Kodari Road would give strategic advantages to the Chinese Communists, [King] Mahendra argued that the highway would be *strictly an economic link*; his aim was to revive the *ancient trade route* to Tibet and restore for Nepal her onetime status as entrepot for the Central Asian hinterland . . .
>
> . . . "The Chinese will get everything from us," said a [less optimistic] trader at Barabise, a center about 12 miles south of the border at Kodari. "They are already getting a lot of our rice, maize and other grains, tobacco, cigarettes, sugar, matchboxes, oil, soap, biscuits, textiles. We get in return only wool and musk, for which there is no market, some salt and sheep. Nepal can't buy a thing she needs. The benefit is all theirs." (*New York Times* 1964, my emphases)

The dynamics of the establishment, disruption, and subsequent rerouting of trade routes is hardly a fluid process. Traders' memories of 1962 often involve the violence of the sudden upheaval of daily life, such as the despair of having to pack up and leave a family shop in Yatung, their concerns over the buildup of armed forces in towns along the borders, and the frustration with the uneven distribution of commodities, as outlined by the quote above. So it was not surprising that when the Sino-Indian border reopened after forty-four years, a ghost story that was wrought by the geopolitical tensions and violence of the 1962 war floated back up to the surface.

Some time prior to the 2006 reopening of the Tibet–Sikkim trade route through Nathu-la, stories of the mysterious "Border Baba"[12] began to reemerge as journalists searched for interesting angles to supplement their reports on the history of the Sino-Indian border region. The stories, in brief, go something like this. On October 4, 1968, Sepoy Harbhajan Singh, a member of the Twenty-Third Battalion of the Punjab Regiment and stationed on Nathu-la, fell into a stream and drowned while leading a mule caravan along the ridges of the mountainous borderlands. Soon after his death, members of the Indian army began seeing his ghost appear in their dreams, where he was mounted on a gray horse. The Baba's sheets were supposedly left crumpled each morning, and some army members who drifted off to sleep while on duty claimed to be awakened after being slapped by him. Today, troops still transport his belongings from Sikkim to his village in Punjab for his annual leave on September 14, and there is a small shrine near the border dedicated to the Baba. According to current members of the Indian army, soldiers on both the Indian and Chinese sides of Nathu-la and Jelep-la claim that the Baba will give them seventy-two hours' warning if any "untoward incident" is to take place on the border (S. Thomas 2002).

A newspaper article concerning the Border Baba, printed around the time of the reopening of Nathu-la in 2006, quotes his sister-in-law, Bibi Satya Kaur, as saying: "Baba was of the belief that the two nations should encourage harmony between them, however, he was opposed to giving a 'free pass' to any country . . . When a road was sought by the Chinese, Baba refused, saying this is not to be given, contending any incident can take place" (Robin 2006). Similarly, Balwinder Singh, Baba's nephew, has said that even "the soldiers on the other side say that they've seen Babaji on a grey horse. He calls for maintaining love and peace with Chinese, but advises on not giving away anything."

Originally, I did not wish to attribute too much meaning to the various supernatural stories I heard during my fieldwork in the Himalayas (yetis were most common); plus, many of the people with whom I spoke mentioned the Border Baba story cautiously, and not without some sarcasm. However, there is something that might be said about such narratives expressing uneasy encounters in border regions. In this case, they serve as a reminder of the history of political violence that has affected communities spanning the borders of India and China, bringing to the forefront something much more sinister than the one-sided celebratory story of reopening the border for free trade in 2006. While the news about reopening the pass was saturating the media, the Baba's family demonstrated their wariness of stronger Indian ties with China, and the army members' ghost stories drew attention to the difficulties of life for troops in high-altitude regions.

The story of the Border Baba is a reminder that examples from everyday life in the borderlands (be it the worry that China is "flooding" India with goods, or a more literal flooding by dam-building and environmental destruction in Sikkim) reveal complicated relationships with the often intangible but still very present state. Perhaps we need to take ghosts more seriously, for "to be haunted is to be tied to social and historical effects" (Gordon 1997: 190). When the Baba stories begin to emerge again in 2006, he arrives bearing several warnings, one of which is a sharp slap that will quickly awaken any soldier who is not paying attention to the potential dangers that lurk—and perhaps have always done so—in the Sino-Indian borderlands.

Of course, to talk about the past is never to replicate exactly what happened. It is more about "seiz[ing] hold of a memory as it flashes up at a moment of danger" (Taussig 2004: 368). This notion of "danger" is apt here, for despite all the talk of the Baba encouraging harmony between the two states, there might be a more ominous reason he is evoked. His simultaneously protective and harbinger-like presence (or nonpresence, for that matter) articulates a fearful discourse of the potentialities of the state—the fear of China and fear of India, and the fear of future violence that might burst forth. The stories of the ghostly Border Baba merge the violence of the past into the present and future by providing reminders of the 1962 border war ("Be careful, it could happen again!") and even the military home-leave as a vestige of British colonial rule. In so doing, they confirm the hegemonic presence of nation-states and borders. "Death mediates the spirit of the state with the body of the people" (Taussig 1997: 103).

At the same time, the Baba's haunting presence disturbs the apparent seamlessness of free trade at Nathu-la between India and China, at least from the Indian side, for he has the power to refuse road-building projects or to warn the local inhabitants of any impending danger. His presence also disrupts the ostensibly rational, nonsuperstitious nature of the Indian and the Chinese states (Yeh 2009). The Baba can almost be seen as an embodiment of the paradox of the border itself, separating while uniting, disrupting and at the same time smoothing over; he is a reminder of potential border cleavages, for he is "calling for harmony, but never giving anything away." A trip to the Nathu-la border and the Baba's shrine is a popular destination for Indian tourists; at this high-altitude monument, his story is told and retold, and his room and bed are put on display for visitors. In these border performances, the Baba's presence is reaffirmed, stoking the "fear of death's contagiousness" in an area marked by potential danger (Rafael 1997: 282). The Baba story thus transcends a single historical event, for with every retelling, the antagonisms, cleavages, and disruptions of the past are regenerated—and stand restlessly next to the reopening of the China-India border for unhindered, friendly trade through a route that has long been checkered by disruptions and diversions.

Specific moments of the history of trade in the Himalayas like these can be used as a basis for analyzing later spatiotemporal shifts in trading relations. Of special importance are the uneven experiences of "opening" and "closing" routes and borders on the part of big and small traders, for these are uncannily similar to what traders experienced with the reopening of Nathu-la in 2006, and the remembrances of war and disruption that get stirred up in the wake of this reopening. Although the most powerful economic players changed between the late eighteenth century and now, the movement of traders remained dynamic and resilient, both thwarting and following, fixing and transgressing hegemonic geographies that controlled the flows of goods across borders. I explore this curious connection between commodities, people, and places in more detail in the next chapter. Traders' everyday experiences with their products—in this specific case, wool—can illustrate how commodity networks have been transformed geographically over time. Older family connections through wool have begun to intersect with newly established (and arguably more successful) Chinese networks dealing with goods such as kitchen appliances and mobile phones. Yet as diversions continue apace, some members of a younger generation of traders are involved in once again revitalizing Tibetan wool for both local and international markets.

CHAPTER TWO

From Loom to Machine

> The world of commodities would have no reality without moorings.
> **Henri Lefebvre,** *The Production of Space*

The World of Commodities

In a book based on the memoirs of Newar merchants who conducted business between Lhasa, Kathmandu, and Kalimpong in the first half of the twentieth century, Kamal Tuladhar writes of his family's shop in Lhasa: "English woolens, Japanese velvet, Chinese silk, Nepalese cottons, and Indian brocade . . . filled the shelves. Coral was imported from Italy, turquoise from Iran, and brick tea from Shanghai . . . Mongolians and Tajiks brought silk to Lhasa, Bhutanese brought rice to barter for silk, Golok nomads brought wool, Amdowas brought Chinese silver coins and gold dust, and Khampas brought brick tea to trade for textiles" (Tuladhar 2004: 22; 69–70).

If we identify this myriad of commodities in the Lhasan marketplace as a material reflection of the social and economic networks linking Lhasa with other parts of the globe, then perhaps we may begin to systematically connect small pieces of "everyday" material culture—turquoise or wool, DVDs or washing machines—with more abstract processes of regional and global economic change. In the Tibetan marketplace, goods involve increasingly complex production and distribution channels. Contemporary brocade jackets may be made of silk brocade from Varanasi, synthetic wool from Xinjiang, and buttons from Kathmandu; perhaps they are assembled and tailored in Lhasa, and sold in Germany. That these networks are opaque evokes Marx's injunction in his writings on the commodity, to peer beneath the surface sheen of the "mysterious" finished product at the web of social relations actually involved in its production, distribution, and marketing (Marx 1990: 168).

Although material culture study, with its focus on the creation of social meanings intertwined with the history, archaeology, and anthropology of things, is rarely recognized a discipline in and of itself, it is perhaps better situated as an "intervention within and between disciplines" (Buchli 2002: 13). This capacity of the study of things to be situated cross-disciplinarily—and indeed, cross-regionally—makes it an appealing vantage point from which to investigate processual changes in the broader political economy of the Lhasa–Kalimpong trade route.

But what does it mean for some*thing* to come from some*where*? How might we investigate the social and economic processes that serve to link places such as Lhasa or Guangzhou with things such as a yak tail flywhisk or a Nokia-esque mobile phone? Who has the authority to create and sell accounts of the authenticity of objects that are at the same time geographic accounts of place? I argue that when the economic geography of the trading landscape is transformed, so are peoples' lived experiences of production, distribution, and consumption; yet at the same time, changes in the means of production redistribute place-based narratives of commodities in uneven and sometimes contradictory ways. In the pages to follow, I focus on traders' and merchants' experiences with their goods to demonstrate how personal narratives of the properties, spatial origins, and trajectories of commodities can serve as examples of establishing or fixing certain kinds of geographies against others.

Commodities and the Social Lives of Things

For decades, scholars have remained divided over how to weigh the relative importance of production, exchange, and consumption in the processes of trade and capitalist accumulation. Two different but related trajectories have attempted to investigate the interconnections between these processes: models examining the "social life" or "biography" of commodities in order to trace their paths through varying phases of production, exchange, and consumption; and models of commodity chains tracing production and consumption processes across national borders.

Anthropological studies of the "lives" of objects have demonstrated that social relationships are necessarily involved in the acquisition, exchange, distribution, and marketing of goods; and research on commodity chains (such as Sidney Mintz's oft-cited study of the cultural history of sugar in the Caribbean [1985] or Theodore Bestor's work on the global sushi market [2001]) have effectively shown how active participants in these chains are both locally and globally situated (Appadurai 1986; Kopytoff 1986; Myers 2001; Thomas 1991; see also Gereffi and Korzeniewicz 1994; Haugerud et al. 2000; Humphrey and Hugh-Jones 1992; Wallerstein 2000). However, as mentioned at the beginning of this book, tracing an object's journey—along with its multiple social connections—from its moment of production to its moment of consumption remains a lofty but unrealizable goal. I focus instead on the importance of traders as intermediaries or mediators in influencing the travels of, and the control over the knowledge of, commodities. Intermediaries such as traders often engage in "the management of meaning," acting as crucial nodes in the paths of things (White 2000). This involves negotiating and directing the authenticity, value, and political significance of objects for exchange, thus restructuring or even fusing knowledge gaps that open up in the spaces between producers and consumers—gaps that often expand and contract due to geographical changes in production processes (Harris 2007).

The role of intermediaries or "cultural brokers" in trade transactions has been addressed most often, however, from the perspective of art markets or dealers (Graburn 1976; Plattner 1996; Steiner 1994), but other studies emphasize the role of the intermediary in long-distance or transnational trade by highlighting their experiences of geographical changes to their cross-border trade networks and practices (Cohen 1969; Hansen 2000; Nordstrom 2007; White 2000; Ziegler 2004). But I agree with Ian Cook and his coauthors that many studies thus far on the geographies of commodities have been politically weak, needing a stronger commitment to defetishize and expose the uneven geographies of the production, distribution, and consumption of things (Cook et al. 2004, 2007).

The commodity itself, as Marx has shown, has no inherent meaning in and of itself; rather, the use and exchange of objects has been historically shaped, so that the same object might have radically different meanings in different contexts. Although there have been critiques of the Marxian approach to the commodity (e.g., Miller 1995), I find Marx's angle critical in paying specific attention to how exchange values and meanings attributed to commodities are both connected to and diverted from places or ideas of place. For instance, certain sheep wool products, no matter what their true origins might be, are seen as tied to a specific kind of Tibetan landscape, one that is "pure" and detached from the current reality of Chinese production, ownership, and development. Keeping in mind the claim that the commodity form obscures original social relations and production (and, accordingly, geographical origins), other kinds of social and geographical connections are reinserted into commodities in complex ways, and are always caught up in political struggles. For instance, an Indian friend told me that during the start of the 2001 U.S. invasion of Afghanistan, the prices of dried fruits such as apricots and dates skyrocketed in Delhi. Although traders had historically brought dried fruit to India from Afghanistan, these particular dried fruits were actually from California. This use of the cultural and historical link between dried fruits, Afghanistan, and India for purposes of profit and/or politics more or less creates a diversion back to the "original" homeland of dried fruit, obscuring the newer geo-economic paths of the goods.

Where, then, might it be appropriate to begin an inquiry into the tangled linkages between places, people, and things? In the words of Henri Lefebvre, it is "in the unmysterious depths of everyday life" where we ought to go to study spatial practices. Lefebvre and others have suggested that social structures and spatial meanings can be extrapolated from detailed analyses of everyday practices. Examining seemingly trivial activities, such as walking along city streets or the routine use of material goods in the home, reveals the varied uses and construction of different kinds of social places—public, private, class, gendered, and so on—as well as social struggles over the control of territory or the unequal distribution of resources. Along with increases in technological progress and the mass manufacturing of consumer goods comes a simultaneous increase in the physical and emotional distance between people and the production of things (Marx 1867; Simmel 1907).

Because one of the goals of this book is to better connect the ethnographic study of things and everyday practices with the more abstract theories of mobility and fixity described earlier, one starting point, perhaps, is to recall Lefebvre's notes on the commodity in *The Production of Space*. I quote it here in full:

> As for the commodity in general, it is obvious that kilograms of sugar, sacks of coffee beans and metres of fabric cannot do duty as the material underpinning of its existence. The stores and warehouses where these things are kept, where they wait, the ships, trains and trucks that transport them—and hence the routes used—have also to be taken into account. Furthermore, having considered all these objects individually, one still has not properly apprehended the material underpinning of the world of commodities. Nor do such notions as "channel", derived from information theory, or "repertoire", help us define such an ensemble of objects. The same goes for the idea of "flows". It has to be remembered that these objects constitute relatively determinate networks or chains of exchange within a space. The world of commodities would have no "reality" without such moorings or points of insertion, or without their existing as an ensemble. The same may be said of banks and banking-networks vis-à-vis the capital market and money transfers, and hence vis-à-vis the comparison and balancing of profits and the distribution of surplus value. (Lefebvre 1991: 403)

A bag of barley flour brought across the borders of Tibet and Nepal is not disconnected from the people who transport it, nor from "moorings" such as warehouses or the mules that carry the goods across the Himalayas. The idea is to consider not simply commodities in and of themselves, but also the social networks that comprise and connect the production, exchange, and consumption of commodities. "Commodities themselves cannot go to market and perform exchanges in their own right. We must, therefore, have recourse to their guardians, who are the possessors of commodities" (Marx 1990: 178). And yet, even the possessors of commodities do not always go to market by their own choice. "Things lie," wrote Lefebvre, ". . . they lie in order to conceal their origin, namely social labour, they tend to set themselves up as absolutes. Products and the circuits they establish (in space) are fetishized and so become more 'real' than reality itself—that is, the productive activity itself, which they thus take over" (Lefebvre 1991: 81). The challenge is to show *how* the various values and meanings inserted in objects by traders are related to larger economic processes. How can we look at the productive aspects of the journeys of commodities together with their "moorings" in order to examine the history of changes along the Lhasa–Kalimpong trade route? How might studying the links between commodities and place increase our understanding of uneven development?

As I mentioned earlier, although a trade route may appear on a map to be a line linking Town A and Town B, the actual "on the ground" experience of trade rarely follows that simple path. The activities that occur in between (and beyond) A and B, in effect enabling a spatial change in the trajectory of the goods—calling suppliers on mobile phones, loading crates of electronics onto container ships, leading mules over the Himalayas, driving trucks—are productive, and are often ignored in the literature on trade.

In the next section, I focus on changes in the materials, styles, and production methods of one very specific item: the Tibetan women's apron (*pang gdan*). At the regional and global level, Lhasa is becoming increasingly connected to successful inland Chinese manufacturing networks with fast production turnovers; however, at the "everyday" level—such as the meaning the apron may have for individuals, or the marketing of the apron for tourists and middle-class Tibetans—there are both widening disjunctions between different sectors of social and economic life and countermovements to reattach earlier, pre-1950s spatial values to objects.

Paths of the *Pang Gdan*

One rather unassuming but characteristically "Tibetan" object of material culture is the *pang gdan*. It is the striped apron worn by Tibetan and Sherpa women in many parts of the Himalayan region (in general, non-Tibetan and non-Sherpa women do not wear these aprons). For the purposes and scope of this chapter, I will not inquire into the folklore, regional styles, names of color schemes, or specific weaving techniques of the *pang gdan* in great detail.[1] Very briefly, however, the *pang gdan* is made of three alternating panels of colorful striped cloth and is often—though not always—worn over the chuba, the Tibetan dress, to indicate a woman's married status. In Tibet, *pang gdan* have traditionally been woven by the women of the household on horizontal wooden looms and made of *shad ma* (goat's hair wool), into a kind of *snam bu* (a heavy woolen cloth), for which the yarn was carded, spun, and dyed with vegetable dyes from India such as madder and indigo. There were different kinds of aprons made from a variety of materials; generally, aristocratic women had more access to aprons made of softer and finer wool like *shad ma*. Bright chemical dyes were first introduced to Tibet in the early 1900s (Snellgrove and Richardson 1968: 235). In the 1950s, these became extremely popular for aprons due to the increased availability of dyes from Europe and India, as well as to the fact that the chemical coloring process proved at least three times quicker than the vegetable process (von Fürer-Haimendorf 1975: 79). By the time Tibet was considered an official part of China in the 1950s, 70 percent of trade between Tibet and India had been effectively bought out by the Chinese state; the Sino-Indian War of 1962 resulted in the sealing of the border passes between China and India, and the Cultural Revolution slowed trade to a trickle (Shakya 1999: 115). More than twenty years after the start of Chinese reforms in 1978, the "Develop the West" campaign spurred massive development and labor migration toward the rural western provinces of China, including Xinjiang and the TAR. It is against this backdrop that our contemporary *pang gdan* story unfolds.

Over the past twenty years or so, a noticeable difference has emerged between apron materials and styles worn by women in the burgeoning urban areas of Tibet and those living in areas further away from major cities and towns. One elderly *snam bu pang gdan* trader from Lhoka noted that more and more urban women purchased synthetic *pang gdan* from retail stores, while those in rural areas continued to wear the homespun woolen ones. "No one," he remarked, "wears wool *pang gdan* in the city anymore." Today, Gonggar County in Lhokha (the southeastern prefecture of Tibet adjacent to Lhasa; see figure 11) remains one of the major seats of wool *pang gdan* weaving, even though synthetic materials are increasingly used. Some nomads in this region near Yamdrok Lake still travel to local villages to sell bales of sheep wool to village apron weavers (generally women), who then use the wool to weave rolls of striped apron cloth (figure 12). Their (mostly male) family members travel to Lhasa during the

FIGURE 11. Map showing Lhasa and Lhokha prefectures in the Tibet Autonomous Region. Keithonearth, http://en.wikipedia.org/wiki/File:Shannan_map.svg, Creative Commons Attribution—Share Alike 3.0 Unported license.

winter and between the planting and harvesting seasons to sell large rolls of the woolen apron material, as well as rolls of cloth featuring a tie-dyed pattern (*thig ma*), often used as decorative trimming on the bottoms of chubas.[2]

Yet, as another woolen cloth trader from Lhokha puts it:

> People in Lhasa don't buy *snam bu pang gdan* as much anymore. It is much easier and cheaper to get Chinese silk now to make aprons. Amdowas and Khampas will buy the *snam bu*, and even foreign and Chinese tourists, but not those from Lhasa. Lhasa people don't know how to make things out of this material anymore. Traders used to sell the cloth inside Meru Nyingba [a monastery on the Barkor behind the Jokhang, the main temple in Lhasa], and back in those days, they used to tell you exactly which items were good and bad. Monks used to purchase the cloth, and they would pay the merchants back in tea later. They used to line the rolls of cloth up in terms of its quality. These days you can't tell which cloth is good and bad. There's more negotiation, and people pay immediately.

FIGURE 12. A *snam bu pang gdan* roll for sale behind the Jokhang in Lhasa, 2006. Tina Harris.

FIGURE 13. Thinner, silkier *pang gdan* thread from Chengdu and new apron design patterns. Tina Harris.

In statements made by other traders of woolen *pang gdan* cloth—who, because of their role as intermediaries, are in a good position to witness changes in both consumption and production—the introduction of this new "silk" apron material from China was brought up time and time again. As recently as five years ago, a shinier, silkier, synthetic thread began to be brought to Lhasa in bulk from Fujian and Sichuan provinces. Although synthetic threads for making *pang gdan* have been brought to Lhasa from places in India with substantial Tibetan communities such as Mussoorie for decades, one *pang gdan* weaver said that in comparison to the new Chinese thread, the Indian materials just weren't as good. The aprons made with Indian thread fell apart quickly, and they were difficult to wash by hand. Most shopkeepers and retailers call the new synthetic apron thread "silk" (*gru'u tsi*) in order to associate it with the higher-value material, but it is in actuality a rayon mix. Unlike the thicker *shad ma*, it can be used to weave extremely narrow and precise stripes, and the finished product ends up looking much sleeker and shinier than the wool *pang gdan* from Lhokha.

While the recent introduction of this new silky *pang gdan* thread may seem like a minor development, it both reflects and reinforces a number of crucial shifts concerning, on one level, the regional economy of Tibet and, on another level, the cultural meanings and value of material objects. From a geographical standpoint, it represents the transfer of Lhasa's dependence for the production of apron materials from less-developed areas in the TAR and cities in India with Tibetan populations to synthetic textile manufacturing centers in inland China. Despite the fact that Lhasa borders Lhokha, over the past five years, faster and cheaper access to synthetic apron materials from much farther away—cities in Sichuan or Fujian provinces—has led to a proliferation of synthetic aprons manufactured in Lhasa, a sharp decrease in the price of aprons, a diversification in the levels of apron quality, and competing apron styles (such as those that incorporate silver and gold threads or new arrangements of stripes). In 2005, for instance, what one workshop owner considered the best-quality handwoven aprons were sold at wholesale prices for approximately 75–80 renminbi (RMB, US$9.00–10.00). Medium-quality aprons were about 50 RMB ($6.00), and low-quality aprons 25 RMB ($3.00). Poor-quality machine-made aprons were sold at retail prices for about 20 RMB ($2.50). Compare these prices with the best-quality handwoven aprons from 2002, which went for approximately 200 RMB ($24.00) on the retail market.

Although fewer and fewer *snam bu pang gdan* from Lhokha are sold to urban Lhasans, they are still occasionally purchased by nomads and pilgrims visiting the city from Amdo and Kham, or by tourists, who traders say use the cloth for decorative purposes—for example, as runners on dining-room tables. From a spatial perspective, then, the trading landscape has altered. While some trading links have expanded and strengthened, others have contracted and weakened. Whereas the apron-making workshops in Lhasa have developed strong connections with and are now quite dependent on materials from industrialized locales like Chengdu in Sichuan, the Lhokha weavers are losing their consumers in Lhasa and must rely on those from other nonurban areas and the occasional tourist.

The access to new apron materials is also related to the increase in Han (in particular, Sichuanese) migrant workers in Tibet over the past five years in the wake of such changes as the campaigns to develop the western regions of China, the easier granting of business licenses for outsiders to work in Tibet, and the relaxing of regulations related to the household registration system (Cn. *hukou*) (Hu 2004). Migrant workers often maintain and reinforce strong obligatory connections (Cn. *guanxi*) with acquaintances and family members from their hometowns (Cn. *laoxiang*), many of whom have access to materials from manufacturers and factories in inland China. On the whole, Tibetans with adequate Chinese-language skills and knowledge of the workings of these migrant business networks have gained access to synthetic thread distributors, whereas those who remain tied to older trading networks are finding it increasingly difficult to keep up with the pace of information and new textile supplies from inland China. With the transformation of the economic landscape of Tibet and its incorporation into the Chinese state, Lhasa has shifted spatially from a regional center into a national margin, losing its "position in the Tibetan cultural imagination as a center of all things Tibetan" (Yeh 2003: 24). While this shift in cultural or geographical imagination may be true among Tibetans in the TAR, it is not quite the same for many living outside China, which we will see some evidence of later on.

Not only do the new apron-making materials illuminate economic transformations on a regional level, they also convey aesthetic and class differences through the various meanings ascribed to them. Prior to the start of the Tibetan New Year holiday, I accompanied some friends on a shopping expedition to purchase new aprons to match their brocade chubas. My friends pointed out the brighter-colored aprons to me as "*har drags sha*" ("too gaudy"); instead, they picked out aprons with thin stripes in subtle colors such as gray and dark blue. In the shops, rows of shiny new aprons were prominently displayed on the mannequins in the front windows. The older aprons made out of Indian synthetic thread were not on show for customers; on the contrary, pieces of them were used to wipe and clean glass countertops. One retailer kindly presented me with a couple of these aprons for the express purpose of cutting them up and sewing them into change purses as gifts for my friends. The woolen *snam bu* aprons were to be found only in individual rolls or displayed in "antique" shops for tourists. When asked what they thought distinguished their clothing styles from those of women outside of Lhasa, a few Lhasan women explained to me: "You can tell immediately if a woman is from the village or the city—only the villagers will wear the woolen *pang gdan* with bigger and brighter stripes." One elderly man who used to be involved with the wool trade in India noted: "The *pang gdan* stripes in Lhasa have gotten much smaller and less bright. Only older, traditional people wear the colorful *pang gdan* with the big stripes."

Uneven production processes are directly linked to these narratives of geographical separation. As some long-standing *pang gdan* networks between urban Lhasa and rural areas of the TAR begin to weaken, the bright woolen aprons come to signify an older, traditional, village lifestyle, while the thinner-striped shiny synthetic aprons increasingly represent a more urban lifestyle. It is important to note that here, the social and spatial gap between "modern Lhasans" and "traditional villagers" is not simply reflected in the wearing of the *pang gdan*; it is at the same time produced and reinforced by the people who purchase them, wear them, and talk about them. And, as outlined further below, these place-based meanings and information about the products can be further produced and reinforced by retailers in order to market the *pang gdan* to more diverse audiences.

Handmade versus Machine-Made Aprons

In central Lhasa, a *pang gdan* workshop is set up inside a simple two-story building, indistinguishable from any other gray building along the busy main road. The workshop is extremely clean but cramped, and four horizontal wooden looms are set up facing each other in a square in each of the two rooms on the first floor. The twenty-odd weavers are mostly young women (with the exception of one man, the brunt of much teasing) in their late teens or early twenties; nearly all the workers hail from Lhoka. Migmar, the Tibetan owner of the workshop, reminisces about the brisk business he experienced when he first opened his workshop in 2000: his aprons were selling at reasonable rates, and his profits increased once he changed his thread supplier from an Indian-based thread manufacturer to one of the Chinese "silk" suppliers from Sichuan.

In 2002, one of his business partners took some sample aprons to a textile factory in Sichuan province and showed the manufacturers how Tibetan aprons were woven. A year later, when the factory in Sichuan developed a machine that was designed to mass-produce *pang gdan*, the business partner left Migmar's workshop to open his own *pang gdan* retail shop and began selling the machine-made aprons that were manufactured in the Chengdu factory and shipped overland to Lhasa. Migmar, remaining at the handloom workshop, found himself hard-pressed to match his former partner's pace and much lower cost of production. At first, Migmar stated, his customers could distinguish between his own handmade *pang gdan* and his former partner's machine-made ones, which were beginning to appear in shops and vendor stalls in the main marketplace, but after a year or so, the machine-made techniques improved so that it became extremely difficult to differentiate between the two upon first glance.

Although *pang gdan* are still woven on wooden handlooms in Lhasa in small workshops such as the one described above and in many village areas in Tibet, handmade aprons have become much more difficult to locate in retail shops around the city. Given that it takes one worker three days on average to weave one three-paneled apron on a wooden loom, it turns out to be much more profitable to transport hundreds of machine-made aprons from Sichuan to Lhasa

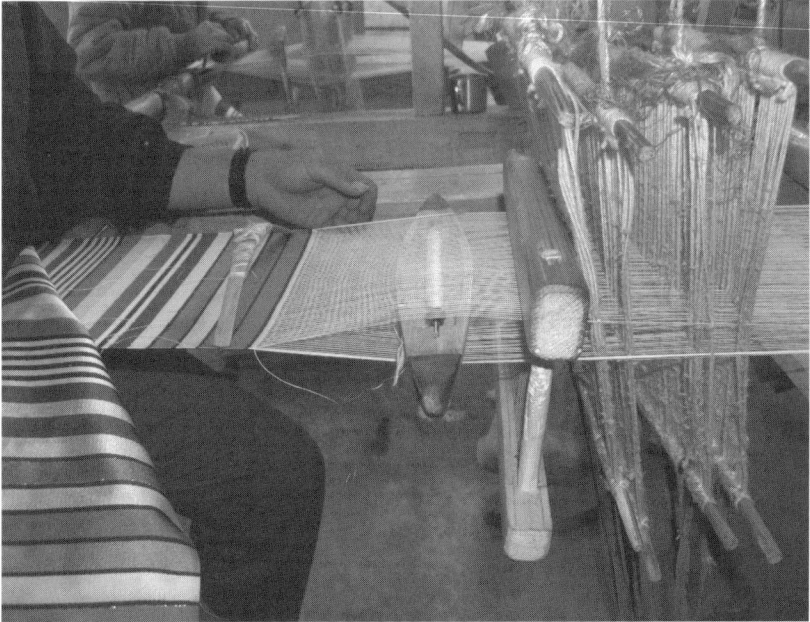

FIGURE 14. Weaving aprons with the new "silk" thread on a handloom. Tina Harris.

to sell in that same period of time. The owners of the handloom workshops are finding that their handmade products simply cannot compete with the cheaper machine-made *pang gdan* from much farther away. This combination of improved *pang gdan*–making technology and better access to economic networks trumps geographical distance, a classic example of how capital will overcome spatial barriers such as mountains or long distances, transforming some social infrastructures at the expense of others (Harvey 1999: 399). Globalization, often either touted as a force that brings the world closer together through new technologies or demonized as wreaking environmental and social havoc, is fraught with inequality, and these contradictions of capitalist expansion take on acute spatial expressions, where some regions benefit and others do not (Katz 2001; Smith 1990). These inequalities in production (or at least the changing relationships of production in the case of just one commodity) reveal a reworking of the geographical imagination that is not coherent, nor does it accelerate in a linear way. It is uneven, not only in the way that it creates "progress" in some places and leaves others untouched, but also in the way that it redistributes place-based narratives in what are often messy and contradictory ways.

To exemplify, both Han and Tibetan *pang gdan* merchants and consumers have stated that although it has become difficult to tell the difference between hand- and machine-made *pang gdan* at first glance, the former are still of considerably higher quality since the threads are more tightly woven and the cloth is softer as a result. In a chuba retail shop in Lhasa, one proprietor showed me the label on a machine-made *pang gdan* from inland China. But the label boasted a Tibetan brand name and a picture of a wooden loom, suggesting that the apron was instead handmade in the TAR. Because intermediaries such as marketers and merchants are often familiar with the geographies of both production and consumption—that is, they know both where the aprons are produced or procured and where the consumers reside—they configure and reconfigure spatial and cultural relationships among the object, its producers, and its consumers. This is done in order to sell their products and obtain a monopoly price—for instance, when a *pang gdan* that is in reality machine-made outside of the TAR is advertised as "handwoven in Tibet."

Perhaps it is true that it is now easier for goods that were once linked to a specific region to be produced or procured anywhere, by anybody (Cohen 2000). To take one example, because of factors such as advanced refrigeration technologies, quicker modes of transport, and overfishing in Japanese waters, the bluefin tuna used for sushi (long associated with Japan) is now fished off the coast of Maine, shipped to Japan's wholesale fish market for auctioning, and then once again sent to the United States for sale in retail markets and high-end sushi restaurants (Bestor 2001). Similarly, the increasing integration of Lhasa's *pang gdan* economy with urban manufacturing areas in inland China has led to the separation of the *pang gdan* from its place of origin. And yet, it is precisely this geographical disjunction that makes it possible for retailers to generate stories and meanings about the objects in an attempt to renegotiate assumptions about their "authenticity," adding value. For many consumers, willingness to pay a high price for a commodity depends on the perception that it originated in an "exotic" locale (Miller 1987) and was produced by "natives." Another poignant example stems from the irony and shock of the western media when workers in a factory in Guangdong, China, discovered that they were producing flags ordered by the Tibetan government-in-exile representing an independent Tibet ("workers said they thought they were just producing colorful flags and did not realize their meaning"; BBC News Online 2008). So when fish used in Japanese sushi is discovered to have come from the coast of Maine, or if an apron is made not by Tibetan nomads but by factory workers in Fujian, diverse stories are produced about the commodities and their associated geographies, reconfiguring existing notions of social and economic networks. And yet these

stories are not always new stories. Sometimes they are older or regenerated stories imposed upon new geographies of production, such as with the label with
the picture of a wooden loom, and in the narrative below.

"Revitalizing" the Apron: New Meanings, New Geographies

As new infrastructural links such as increased flights from Beijing and the
Qinghai–Tibet railroad bring different kinds of clientele to Lhasa, retailers
find that they must market their goods accordingly, selling their wares in increasingly diverse ways to match the demands of this new flood of consumers.
Because *pang gdan* serve as more or less utilitarian clothing items for many
women in Tibet, and because non-Tibetans will rarely purchase a *pang gdan*
to be worn, merchants have lifted the design of the *pang gdan* from the apron
itself and used it to create diversified products geared toward tourists. Like "Yak
Brand" yogurt or illustrations of Tibetan opera masks on the labels of bottles
of spring water, the design of the *pang gdan* has also been appropriated as a
symbol of distinct "Tibetan-ness." One can now find aprons that have been
made into pillowcases and bags, wall hangings made from old discarded *snam
bu* aprons, and carpets woven to look like giant *pang gdan*.

   With regard to the contemporary transformation of consumerism in Lhasa,
it is important to note that the representation and creation of the "Tibetan-
ness" of objects is a process that is not solely in the hands of consumers. In a
paper on the production of Tibetan carpets in Nepal, Eric McGuckin (1997) has
suggested that scholars should pay attention not only to the criteria by which
different consumers construct the authenticity of Tibetan goods, but also to
the more indigenous forms of authenticity that emerge out of such discourses.
Ethnographic research on cultural ideas of authenticity with regard to material
culture has been helpful in examining how power over "objective" knowledge
is formed and used (see, for example, Handler 1986; Steiner 1994). But while
authenticity remains an important concept to debate, I have found that the
question of whether a product is "authentic" comes in and out of use at certain
times. For instance, most apron merchants consider machine-made aprons in-
ferior to handmade ones in terms of their composition, but not necessarily inauthentic. In the case of mock brand-name items in Vietnam, for instance, the
authenticity debate sometimes ignores the complexity of the term: "although
the terms *authentic* and *real* often are used interchangeably, they have different
meanings" (Vann 2006: 294). The dialectical processes between sellers, buyers,
and marketers of making something authentic has been aptly deemed "socially

FIGURE 15. Decorative wall hangings made from old *snam bu pang gdan*. Tina Harris.

ordered genuineness," where authenticity changes depending on whose hands the object passes through (Spooner 1986: 225). In this case, once the apron-making machines improved, few people were able tell the difference between machine-made and handmade aprons. The seller's knowledge of changing demands (local and "foreign") from both the producer's side and the consumer's side is crucial in informing the type of cultural, geographical, and aesthetic significance that is attributed to the profiting from the "place" of *pang gdan* products. The following story, focusing on the intermediary role of a seller of *pang gdan* handicrafts, serves to demonstrate how various discourses of authenticity are intertwined to form new *pang gdan* meanings, and how such meanings serve to reattach place to objects.

From Kathmandu's heyday as a hippie destination in the 1960s to its current position as a Himalayan trekking hub, much of the city's economy over the past fifty years has been dependent on tourism. Boudhanath (Boudha), one of the two major Tibetan neighborhoods in Kathmandu, is a part of the city that often attracts foreign tourists interested in studying Tibetan Buddhism. Some of my foreign friends had told me that if I wanted to see some interesting things that were being done with old woolen aprons, I should speak to the proprietor of a textile showroom in Boudha, who collects old aprons and turns them into decorative wall hangings for foreign expats, tourists, and more recently, upscale hotels in Lhasa (figure 15).

When I stepped into the showroom (an open-air enclosure under a corrugated tin roof), Pema greeted me and gestured for me to sit down in front of him, much in the style of a monk giving teachings to a student of Buddhism. Pema is a middle-aged Tibetan entrepreneur, born in Nepal, who has recently made several trips to Lhasa in order to sell decorative household items—such as bags and wall hangings—made out of old, discarded, handwoven *snam bu* aprons (of which there seems to be a surplus now that many women prefer to buy synthetic aprons). Some of the larger items, priced at well over a hundred dollars, are aimed at very wealthy groups of tourists who visit Tibet. When I asked him to describe his trips to Lhasa, he replied:

> **Pema:** You know, now, there is a train coming to Tibet. And of course the train can come. But in the harsh winter, who holds Tibet back on its feet? The Tibetans. The nomads. Because they are the ones who can really survive in Tibet, work in Tibet, they are born in Tibet, and brought up in Tibet. So what we are trying to do is we are trying to support them. How do we support them? By recycling these things [aprons].
>
> **T:** How does it work when you get the old *pang gdan*?
>
> **P:** So what we do is [take] the good ones, we restore it back, we join it back, stitch them back and make it into a big blanket . . . There are many items around in the Lhasa market, and everything is either from China or Nepal or from India. But these aprons have the real energy, which has been woven in Tibet before, in this place. So this is the original Tibetan stuff . . . If I hang this wall hanging up in Tibet in some open place, all these old ladies and old grandpas and all . . . I want to remind them, "You were like this before." And how are we now? We should be like before.

At first I wasn't sure what to make of this; much of the interview was spent shifting uncomfortably on the floor, feeling skeptical about Pema's claims to understand the livelihoods of nomads, being self-conscious about my position as a young foreign female researcher, and wondering if I should mention that I wasn't in Boudha to study Buddhism. The conversation, however, was a rich tangle of different discourses; the "energy" of the past attached to a specific place (Tibet) and people (nomads), and both the place and the people ostensibly linked to his products for sale. Pema's somewhat embellished statements suggest that he is aware that his foreign clients' demand for "authentic" Tibetan products means that the items should be made *in* Tibet *by* Tibetans. The irony is that a high percentage of the very *pang gdan* that Pema uses to make his wall hangings are made in non-Tibetan villages in Nepal. Yet his *pang gdan* hangings are "real" for numerous reasons. First, because some tourists are aware that many items in the Lhasa marketplace are actually manufactured in places outside Tibet, these hangings are seen as more intimately connected to the Tibetan landscape, and therefore, as Pema puts it, the "original stuff." Second, unlike the other goods sold in Lhasa, he says that they provoke a relationship with elderly (and therefore somehow more "traditional") Tibetans. Furthermore, they are handmade with "the real energy" of the past and in contrast with what he considers the rampant consumerism and contamination of Tibetans in the TAR today. Here, the "Tibetan-ness" of the aprons is both created and reinforced by Pema's place-based narrative, various strands of which are taken from his own detached position as an exiled observer of an urbanizing Lhasa, as well as his understanding of what tourists will purchase while visiting the TAR. By putting these recycled apron items on the market, he insists that he is supporting the nomads; curiously, Pema—who lives in Nepal—puts himself in a position of authority to represent what Tibetan nomadic space should look like as well. The value of the goods is based on the production of a pure, pre-industrial aesthetic that is no more: these are secondhand aprons that are no longer in use, yet their "vitality" and "energy" lives on. Pema's narrative of the revitalization of woolen aprons is linked to larger place-based narratives of decline that came up in conversations with other merchants and traders.

Narratives of Decline: The Past as "Natural," the Present as "Fake"

From my field notes in Lhasa on November 30, 2005:

> Excessive hanging out today. I am absolutely full of sweet tea to the extent that dinner with DK and E and K will be a little too much. It was a bit of a tough day— I just had no confidence that I can pull off this project. Went to the teahouse to just hang out with the girls [the staff] and learned that K [in her early 20s] is an orphan with 10 brothers and sisters from Gonggar; she worked at the teahouse for 8 years, where the other girl worked there 13 years. K told me about a TV show where butter sellers cheat by putting a potato into the butter mixture to make it heavier and less pure—a woman went to the police because of this. A funny *rmo lags* [older woman, literally "grandmother"] was there as well—she warned me of the many *rku ma* [thieves, robbers] around town.

The cheese seller says the Hui [Chinese Muslims, one of the fifty-six ethnic groups in China] use his cheese to make makeup! This is something I have heard elsewhere. They use the *phyur zhib* [crumbly or fine, powdery cheese]. Oh—and another rumor about the Chinese buying up tsampa [Tibetan barley flour] also for using in makeup. Can't help but wonder where these ideas come from.

Met a woman, 81 yrs old, she seemed very worried—X thinks she probably knows quite a bit more than she is letting on. She doesn't say a thing about mule trade, and she doesn't want to be recorded, which is understandable. It seems as if there may be something she has witnessed or some episode with a friend or relative that didn't go over so well. She did say a few things of interest: in the past, people could wait for their money or goods for at least over one month, and that people used to trust each other. She seems very skeptical of people today. She said she was old, that she didn't have much time left. "People didn't pay immediately like they do now," she said. Also, "now everybody has an education so you can't trust anyone." And, "back in the old society, the quality of items was much better, the people were very good, business was good. Goods from India were of better quality, you didn't have to pay tax like now."

In my conversations with both old and young traders, I was struck by one prominently recurring theme: that the quality of commodities in the past far surpassed the quality of contemporary items. Not only were manufactured goods considered of better quality, but even the food was more "pure." Everything in general was simply more real, unlike commodities today, which were often described as fake, contaminated, or impure. One woman in her seventies described how *gla rtsi* (musk) and coral used to be brought from Kashmir and sold to *bar mi* (middlemen or intermediaries) in Lhasa, who would then barter these items for butter brought by nomads. "Now all the coral is fake," she lamented. "The quality goes down over the years."[3] Similarly, one elderly Tibetan trader, T. A., who ended up escaping from Lhasa and moving permanently to Kalimpong in 1959, recalled what commodities were like in both Tibet and Kalimpong when he led caravans between those two towns in the 1940s and 1950s:

> '*Go snam* (feltlike wool) was British. It was such good quality, that you could roll it, it would stay together in a roll and you could stand it on its end and it would not fall down. All this stuff was available in Kalimpong . . . Oranges were sweet and tasty; for 1 rupee you could get 100 oranges; things back then, fruit, for instance, was counted in the hundreds, it was so abundant . . . You could eat as much as you would like . . . Now you hardly find them at all. There [were] much

better sheep in Tibet, everything was better . . . Reds were redder, even colors were better. You could get anything in Lhasa! Everything was available! Man can create things . . . he can manufacture . . . but he cannot create natural things. You would think that with more things our life would have improved by now.

Reds were redder, even colors were better. Here, I find that Raymond Williams's study of English historical perspective through pastoral literature in *The Country and the City* provides a window through which to examine traders' narratives concerning the items they dealt with. In the context of traders' experiences in Tibet, there is a common narrative whereby natural, real, or living things are seen to be from a specific time and place—in this case, from rural, pre-1950 Tibet—and items nowadays are considered fake or impure. What Williams calls a "structure of feeling" is persistently created throughout history. Through actual, lived experiences, there is a "felt sense of the quality of life at a particular place and time," reflecting the ideology that things were somehow much better back in the day (Williams 1961: 47). While there has always been a recurring, familiar nostalgia for an ordered and "pure" past, such virtues or notions—sweet abundant oranges, better sheep, vibrant colors—"mean different things at different times, and quite different values are being brought to question" (Williams 1973: 12). The shifting narratives of what makes certain commodities "pure" and "natural" are linked to changing economic contexts and representations of space.

For instance, I asked one Tibetan Muslim woman working in a small jewelry shop in Kathmandu (one of the first shops in Kathmandu to sell jewelry made of stones from Tibet) why she thought things were fake now. She replied: "because of the situation of the country. Before, there were hardly any of these types of shops. Now everywhere you go, you see the same kind of shop. People copy things, there is competition, they have more capital. And there are many fake goods. Natural things have a good high price. So now there are no good natural things, so it is cheap. And our customers like cheap things."

Although an elderly Marwari man with a shop on R C Mintri Road in Kalimpong who used to sell pens, watches, and cloth to the traders going to Tibet thought commodities were actually cheaper in the past, he shared the idea that they were then of better quality. He exclaimed: "The town was so good back then, people made profits, and oh, the goods were so cheap! Seventy-five paise for a meter of diaper cloth, one rupee for three liters of milk, one rupee for two kilograms of atha flour. And oh, the quality! Nothing like it is now, where the daal is mixed with other things, stones and bad-quality daal. The milk was completely pure, not like today where it is mixed with water."

Or take, for example, an Amdowa man living in Kalimpong, who traded in Lhasa between 1956 and 1959 and said the following:

> If you tell this to the young generation, they will never believe me, but burnt cow dung is a much better fuel, the food tastes much better than if it was made with gas. I really enjoyed my time as a trader in Lhasa. In the 1960s, there was no need to grow squash, you could find it anywhere. Now you can hardly get it, and it is so difficult to grow. Before, in Lhasa, everyone's *gshis ka* (personality) was good, even the Chinese from Beijing spoke great Tibetan, Nepalis spoke Tibetan, Marwaris spoke it too. Most people spoke a very good Lhasan dialect of Tibetan, almost everyone used it, even the Chinese. What is it like nowadays? Like there's American money, and American business everywhere. The women and men worked equally, not like here in India, where women stay inside the house all the time. But the women didn't go with the muleteers. They didn't usually travel, but they stayed in the shops.

The notion that natural things are now either fake or contaminated was prevalent in many conversations with traders. Fatima's mother, who owns the shop mentioned above, had this to say about the quality of foodstuffs in the past:

> That's why (older) Tibetans are so strong. Yak butter. They eat all natural food. This time all things are not made of natural food. Egg, meat, everything is not good. These times, many people get sick. Cancer, blood cancer, everything. My mom said, in our time [1940s, 1950s], they made food in a clay pot, eat rice, then meat, then they throw away (the pot). This time, clothes clean? No! Everything is dirty. But their time was not like this time, cancer-shmancer. (My mother's) time was very clean, every time they took baths, but our time is very bad. That time, old people lived a very long time, 90 years, 100 years, 102 years. This time, people live only 20 years, 25, 30 years. My mom, my grandfather, ate all real things. *Thud* [a dish made of butter, sugar, cheese, and barley flour] was very fresh. Real things. Green things, and boiled things.

The contamination of food, the fake goods, the *cancer-shmancer*—all reflect the separation of people from products that were once more knowable, the origins of which were once known. According to Fatima's mother, in historical Tibet there was life in everything—people lived longer. You could trust everyone not to tamper with foodstuffs or raw materials. But how consistent are these memories of a "pure" past? In direct contrast to the stories of how wonderfully natural items were back then, there are many traders who remember dirt and dishonesty: plenty of it.

An elderly trader named Kulbir remembered how there were few vegetables in Lhasa—no tomatoes, no cauliflower—and those that were sold were often rotten. He noted the lack of rice and the fact that you had to be rich to afford mutton. There were stories concerning how the wool trade was plagued by untrustworthy types who were actively working to defile the wool. Kulbir would have to poke the big bales of wool when they arrived in Kalimpong to see if there was anything else in it: "Sometimes they stuff it with sand," he said. "Sometimes they put rotten leather inside." Similarly, I heard one story about a trader who used to bring wool to Kalimpong from northern and western Tibet in the 1940s and 1950s. To make the wool heavier and more profitable, the nomads would dip it in water and urine; when it dried, it became sticky and heavy. In addition, dirt and sand might be mixed in. The trader who told me this story laughed, saying that they weren't exporting wool from Tibet, but "they were exporting Tibetan sand to India!" Furthermore, when the wool reached Kalimpong, some of it would be rotten and useless, so they would have to take time to select the best kind for further export. When I asked the man who told me this story how the traders would pay when this happened, he replied that the traders would trick the nomads in turn, because they would cheat on the weight; as "nomads didn't know much about weighing, they [the traders] would say it was lighter than it was, and all would be relatively equal in the end."

In 1946, an anonymous letter was sent to the British Trade Agent of Yatung, Tibet, and addressed to the secretary of the Foreign Affairs Department, New Delhi. Here, an angry trader writes of the defilement of textile bales on their way to Kalimpong:

> In spite of the fact that quota holders and Commission Agents are not allowed to open the unpacked bales the Commission Agents of the said Syndicate open them and give them for dyeing purposes to different small factories in Bombay. In dyeing the length of the cloth increases by 2 or 3 yards and they (Agents and Commission Agents of the Syndicate) appropriate the excess yardage to themselves. They collaborate in illegal manner with the local dyeing factories and although they have it stamped on the cloth as "Fast Dyeing" actually they put Katcha dyeing [poor quality, non-permanent dye] on it and thus defalcate a large sum of money . . . They in certain cases have taken out certain yards from inside of the pieces and have kept them to themselves there and thus have cheated the Tibetan merchants . . . After opening the textile bales at Bombay they repack the bales and inside the bales they put woolen and many other kinds of goods which come direct to Kalimpong without any hinderance [*sic*] from any of the people

FIGURE 16. Sorting and weighing Tibetan wool, Kalimpong, circa 1930s. Photo by K. C. Pyne, Kalimpong Stores (Kodak).

concerned under the permits given to them as quota for Tibetan Syndicate. (India Office Records 1946)

Further, consider a contemporary account from a trader who would travel to northern Tibet to collect wool from nomads as late as the 1990s:

The villagers already collect their wool themselves. From one village they collect all the wool and they are just waiting for us to trade. And then we visit the village and we talk to the manager or the leader of the village, and then we buy their wool, but the first time it was 40 tons, like 10 trucks (laughs). It's just like there's soooo much wool there. And then we brought all the wool all the way from Changtang to Lhasa, and we have to wash it and get it dry, and there's so many processes involved. Wool trading is really tiring, honestly. And it's really risky as well. The wool itself is pretty light, relatively, so that if you are not careful, people will just like play tricky stuff. There are some nomads that are so wise (laughs). You know, they, when they pack the wool, what do you call it, they screen it and then pack it in a roll. But when they do that, they do it on the ground where there is a lot of dirt and even sand. And a lot of sands are in the wool, just got in the wool, and then the weight gets just so heavy. And they just do it intentionally, purposely, until to weight up a little bit (laughs) so that they can make more money. And it's so risky for us as buyers.

Although these two sets of remembrances—of pure, "real" products versus dirty wool—demonstrate that the recurring theme of "the good old people succeeded by the bad new people" is indeed a "seductive song," they are still of significance (Williams 1973: 83). Whether things were in fact any "cleaner" or "dirtier" in the past is not likely to be determined anytime soon (nor are these attributes necessarily binary opposites), but what I would like to emphasize here is that the attachment of values to objects is directly related to changes in the geographies of production. The remembered "purity" of fruit or wool, for instance, is set against the current influx of cheap goods coming from China, and is therefore located in a very specific time and place: a pre-1951 Tibet or Kalimpong. The distance said to be felt between people and their formerly "knowable" foods or commodities is part of larger historical processes of capitalist expansion. When I asked one man in Tibet whether some people tried to trick traders by making fake products or adding dirt to wool to make it weigh heavier, he simply answered, "Before the Peaceful Liberation, nothing like this happened." The "Peaceful Liberation" (*zhi ba'i bcings 'grol*) is the official term that the Chinese government uses to describe its entry into Tibet in 1951; occasionally Tibetans will use this term cautiously and with a hint of sarcasm. When talking to strangers, I found it extremely difficult to tell the difference between the genuine and sarcastic uses of this phrase. "Before the Peaceful Liberation" or "after the Peaceful Liberation" was often used as a temporal marker, similar to "old society" (*spyi tshogs rnying pa*) and "new society" (*spyi tshogs gsar pa*), which are also used to demarcate a timeframe or political era. In Tibet, the term "old society" (*spyi tshogs rnying pa*) is temporally and politically significant, as it is often used by Tibetans in Tibetan-speaking areas of China to refer to a pre-1951 Tibet. *Spyi tshogs rnying pa* has been used in history textbooks to connote an unchanging, pre-Communist, feudal Tibetan society (Schwartz 1995: 150) but is sometimes used by people simply to mark the time before the Chinese arrived. Of course, Communism did not abruptly "begin" on this date, immediately wiping out what came before (whether feudalism or a "severe social hierarchy"); changes in political structure do not necessarily run parallel to flows of capital. It is no question that the three elite trading families had extremely strong control over nearly all of the wool production, exchange, and profits in the mid-twentieth century, but pre-"liberation" wool and goods can take on meanings that are deliberately detached from these kinds of profit-making narratives and become more closely associated with the economic drive of China today.

Kulbir states the situation more clearly by saying, "At that time [before 1951], the quality of goods was very good. Then after the Chinese came, new Chinese goods added many levels of quality." At this point, his daughter interjects and talks about her own business in Guangzhou: "If you want to buy pens and they say one piece is 10 rupees and you say, 'No, I think that's too expensive,' they will pull out an identical pen and say, 'Ok, this is the 5 rupees quality pen.' They can make it look the same, but there are many levels of quality." The daughter continued to say that the increase in choice was actually a good development, especially for her business. This notion of having a single level of quality in the past—general good quality—is directly linked to social, environmental, and economic developments in the "new society." Take, for instance, what Pema has to say about the purity of the older wool aprons in contrast to what he considers the contaminated wool of today:

> Sixty years ago, the wool itself had very good genes. The environment was very clean, and we had very good animals, raised in a domestic way in Tibet, so there was no chemical pollution or anything. So we compare it with the wool which is now in the twenty-first century and there is lot of difference. There is a lot of dead fibers these days . . . these days what they do is they use lot of chemicals, mixtures, and then sometimes even you find nylon and polyesters involved with the woolen materials. And they even tag them with very good names, like sometimes they say like "Giordano" or whatever—you know, big companies, they say it's woolen. But if you thoroughly check the material and look into it, you can find oil products.

When I asked Pema to tell me about what he knows of the history of the wool trade, I was not aware that when he said he would "start at the beginning," he meant prehistoric beginnings:

> So nomads, they first used the wool and made felt from water and sand. So this was human dress for survival, which could be used at night and day, without tents, or without shelter. I am talking about very early. And when they used the felt, they used to use it around their waist, you know, very thick felt around their waist, and that keeps their kidneys warm. And they sleep under rocks, inside caves, and they didn't have animals, you know, at that time. They used to go like nomads, like hunting, whatever—this was like the stone age time, how the wool evolved in Tibet. After the felt, then came some clothing, then tents. And then slowly, they started to weave carpets.

When this evolutionary handmade aesthetic is emphasized, the tourist (or the anthropologist, in this case) is made to feel connected with an imaginary nomadic weaver in a precapitalist Tibetan rural scene. Like the shirts he says are "tagged with Giordano," many of the "Tibetan" *pang gdan* are from Nepal. Again, control over geographical information—in this instance, understanding that *pang gdan* from Tibet seem more "authentic" than those made in Nepal, even though Sherpas in northern Nepal also weave similar aprons—is important in order to manipulate the object's meanings to fit the agenda of the moment (Steiner 1994). The question, then, is not about authentic versus inauthentic, or pure versus dirty items, or a single level of quality versus varying levels of quality, but about who has the authority to create and sell accounts of authenticity that are, at the same time, geographic accounts of place.

Pema's narrative is reflective of the shift from one market segment to another, where the value of the goods is based on a particular perception of authenticity or pre-industrial aesthetic, and the items are marketed as such in order to gain a monopoly price. Indeed, he meshes Western and local Tibetan environmental discourses and perceptions of the market with his experience of the global tourist market.[4] Wool, according to Pema, is "hampered" or killed due to global "friction": the development of chemical processing, the creation of nation-state boundaries, and Tibetan nomads who embrace the new Chinese consumerism. And his solution to this friction is to market recycled woolen *pang gdan* from Tibet. The following statement, quoted in its entirety, expresses these thematic connections:

There is friction in the planet. Every time there is friction, every time cars are moving, planes moving, and the roads are paved, and it's so hot, and houses are getting built up everywhere, cement is coming up, yeah? And the green planet is getting hampered by all of this . . . So, you know, there is a cause for the wool getting hampered. So what is the cause? The cause is the modern high technology of the oil products. The animals are dying in Tibet, yaks are getting killed, by all these chemicals, trying to inject them. And sometimes there is a big problem, because sometimes they say, "This is Nepal, [*gesturing a border with his hands*] and this is Tibet." "This is Nepal, this is China." And the yaks cannot come and eat grass. Cannot come and eat grass [on the other side of the border] in the winter when there is no grass in Tibet. In olden days, when there is no grass in Tibet, the yaks used to come down to Nepal and eat grass . . . The *drokpas* (nomads) can have more wool. So this is very important. The necessary movement is to recycle the old products and show them to the world, so that they [the nomads] will understand. And to show them the way they used to weave in Tibet, because

they want to have TVs at home, they want to go for karaoke . . . If we can explain
to them, if we can do something for these nomads by explaining, and upgrading
their products, in a foreign market, which is better than their own economic life,
then that's it.

After he gets up to take care of a customer, another Tibetan man in the
shop, perhaps an assistant, noticing me silently admiring the wall hangings,
said, "This is old, hand-dyed, handwoven wool from our culture. *We must keep
it alive* as much as possible, we must save every scrap and use it" (my emphasis). What is significant here is the insistence on revitalizing the wool, keeping
it "alive as much as possible," where values of purity and life are reinserted into
the wool in order to make it "real" again. These discourses do not therefore
expose a clear-cut opposition between cleanliness and impurity. The reconnecting and revitalizing that is taking place is both a spatial and a temporal action,
highly dependent on locating or placing the wool back in Tibet. It is wool that
is not only Tibetan, but woven in Tibet before Tibet became part of China. The
irony, of course, is that these items were not actually made in Tibet, and they
are being sold not to nomads, but to foreign tourists and wealthy Chinese and
Tibetans.[5]

Dead Labor and Living Wool

Given the examples above, the *pang gdan* economy seems to be in the middle
of several transformations with regard to the value consumers place on the
method of production. In tandem with shifts in the geographies of handmade
versus machine-made products, there are changes in what it means for an object to be "handmade" or "machine-made." "Handmade" may very well still signify "old-fashioned" or "traditional" to a majority of Lhasans, but to some new
groups of consumers (especially a growing number of wealthy Chinese tourists
or Tibetans who may have lived for some time in large cities like Beijing), own-
ing a handmade, decorative item from Tibet may actually distinguish one as
"modern" or a well-to-do person who cares about high-quality, unique items.
In some cases, however, the consumption of handmade apron products may
signify a return to, or a refixing of, a "Tibetan-ness" that is felt to have been
distant or missing, particularly among middle-class Tibetans who spent most
of their educational or professional lives in other parts of China.

Changes in the Chinese economy and the rise of a Tibetan middle class in
Lhasa mean that "old" and "traditional" items such as woolen *pang gdan* may

now be seen as valuable for a new reason: they symbolize distance and rarity to consumers who may be used to fast-paced urban life. Pema advertises his woolen items as providing a potential bridge between nomad producers and wealthy consumers, the past and the present, or nature and urban life; paradoxically, however, a sense of distance still needs to be maintained. Sellers will market ordinary goods by making them distant, exotic, and yet somehow accessible at the same time (Simmel 1990). The resulting tension between distance and accessibility is formed with the acute knowledge of a particular market—its history, its producers, and the tastes of its consumers. The very use and display of this new form of *pang gdan,* not as an everyday clothing item but as a piece of household décor, further separates those who can afford these labor-intensive, high-end objects of tradition and those who cannot.

Dawa is a Tibetan businessman in his mid-thirties who returned to Lhasa approximately ten years ago, after finishing his schooling, to work in a retail carpet shop. Part of a growing group of middle-class Tibetan entrepreneurs who speak Chinese, Tibetan, and English, Dawa recently left his job to take advantage of expanding opportunities in the tourism industry in the TAR. He is interested in selling upmarket handicrafts, including items made from woolen aprons. Many of Dawa's peers consist of the (now grown) children of relatively well-to-do Tibetan families who have been sent to inland Chinese boarding schools and universities so that they might gain better employment opportunities in the future (Shakya 2008). While discussing his dream of opening a large handicrafts center in Lhasa, he explained how he wanted to market his goods to new groups of consumers. Dawa remembered how, in 1997, 90 percent of his carpet shop customers had been western tourists, compared to 2006, when his clientele had been at least 50 percent Chinese. There has been a dramatic change in the number of foreign versus domestic tourists visiting the TAR since the 1990s, mostly due to rapid increases in transport connections to and from Beijing, as well as a government-initiated agenda of promoting tourism and leisure culture in order to increase domestic consumption. In the 1990s, foreigners made up the majority of tourists to the TAR. Domestic tourism, on the other hand, increased by 30 percent every year between 2000 and 2005. In 2007, the year after the inauguration of the railroad, the number of domestic tourists came to 3.4 million, far surpassing the 2.6 million population of the entire TAR, while the number of foreign tourists was 338,744 (All China Marketing Research Co. 2008; Murakami 2007). Dawa described how five or ten years earlier, Tibetans would come into the shop, "and when we would tell them the price of our carpets, they were shocked. But now the Tibetans who come to see them are ready to bargain. 'Give me a little bit of a discount and I will buy this carpet,' they say—even though they are paying double the amount of what they are getting in the market!" Dawa was particularly excited about the potential for the growth of middle-class Tibetan consumer markets for handicrafts. "We have to make these things [handicrafts] kind of like a fashion for Tibetans," he said. "The trend is coming."

> People don't know how to make these things anymore—maybe their parents do, but they don't. I think even the younger generation has a problem. We have to make these things kind of like a fashion. I will give you a small example. In Switzerland there are a lot of Tibetan communities. During a couple of years ago, for example, if they wore an apron, the apron colors were really, really bright with these huge stripes. It was kind of like a fashion. But that's what the villagers [in Tibet] were wearing. Now my friends in New York, if they want to wear an apron, they want to find this kind of woven apron. But over here in Lhasa, the trend is totally different. They want to look really modern, wear the [new aprons]. I think it sometimes happens that when you leave your country and when you go to other places, then you feel more Tibetan, then you feel that you need to keep your culture and everything. It was so funny seeing my friends when they got their visas and they decided that they were going to stay in the United States. Everyone started to buy Tibetan dranyen [a traditional Tibetan musical instrument, similar to a lute or a banjo], and everyone started to buy Tibetan rugs and they started to learn how to play the Tibetan dranyen. Now it is kind of like—guitar out, dranyen in (laughs)!

Dawa continued this discussion of the transformations in Tibetan consumption practices by explaining how he was having a conversation with one of the designers for his modern handicrafts center in Tibet about where the workers might live and how the company building might look. In contrast to some existing shops and workshops designed in a Tibetan style and geared toward enticing tourists, Dawa's idea was that he would build sleek, contemporary buildings for the workers, "because I know that nobody can be fooled by a Tibetan-style building when this is a factory with weavers. Most of these weavers actually come from the village. They are so used to seeing Tibetan-style buildings. But when they come and look at the building they feel like working [in], they kind of feel like they are in the twenty-first century. Because I know their kind of mindset. It is not like the Westerners'. Westerners are so fed up with seeing these modern buildings that they look for the Tibetan-style buildings. There is this saying in English: the grass always looks greener on the other side."

Over dinner in Lhasa, Dawa also brought up the importance of revitalizing the older *pang gdan* wool in order for it to be a "trend." He described the wool as being "alive" and suggested that the use of machines in some ways acted to "kill" the wool. He said: "When you have wool, if it is machine carded, the machine doesn't have—it doesn't have a mind, so the fibers are actually broken down, so that is kind of like removing the life of the wool. If it is machine carded, then you just take the life out of the wool . . . you are actually taking the life out of the wool."

The various attempts to revitalize wool for the new middle-class Chinese and Tibetan market also reflects the movement to reconnect or "fix" places and geopolitical or cultural relationships that are considered "lost" or "broken." The idea that wool from pre-1950 Tibet is "alive" and that new, machine-carded wool is "dead" is reminiscent of Marx's notion of "dead labor," whereby products and the means of production (e.g., wool and machines) are seen to be productive in and of themselves. Revitalizing the wool by recycling it and turning it into a handmade product not only involves conjuring back living labor; the wool itself is seen as endowed with living properties (not the laborers who make the woolen handicrafts in the modern workshop) and therefore recognized as having more value. The "real energy" said to exist in the wool is thus more than just a simple opposition to "dead" wool. It is laden with multiple values involving a change of location (labeling it as "Made in Tibet" as opposed to "Made in China" for instance) and a change in temporality ("It was produced like this in the past"). Talking about the "real energy" of the wool is also one way to make visible—for a profit—a Tibet that appears to have become a blind spot amid the larger, China-driven economic shifts of the region. What this does in turn, however, is obfuscate the location of the producers of the object, who may not actually be Tibetan or living in Tibet. Since there are no inherent meanings in objects themselves—rather, the meanings and uses of goods have always been formed by the *people* involved in their production, exchange, and consumption—the values that are attached to wool at its consumption stage can be very different from those attached during production. With the emergence of new geographical paths of capital, these narratives expose different, and occasionally conflicting, valuation systems. For instance, the silky new "urban" aprons signify a modern or sophisticated wearer; new machine-made aprons mean a loss of the old virtues, the traditional way of life, and high-quality production skills; and the older aprons with the bright stripes connote that the Tibetan living in Switzerland identifies herself culturally with Tibet, as someone who is truly Tibetan. Someone who sells high-end handmade Tibetan goods claims the authority to weave place-based narratives about nomads or other producers like apron weavers, transforming machines into Chinese killers and reanimating the dead Tibetan wool for consumers. But it was never the objects themselves that could act in the first place.

Objects and Places

Place-based narratives about objects are of great significance to the study of uneven development, particularly in regard to contemporary globalization processes in western China. The social lives of things rarely take on a linear trajectory and are often unequal, hierarchical, and rather curious in the sense that commodity paths can be diverted through both changing modes of production and creative narratives of place. Instead of treating ethnographic studies of material culture as a separate field to more macro-level studies of political economy, I argue for the integration of the two, since larger-scale transformations in transport networks and the means of production redistribute spatial narratives of things in ways that do not necessarily line up with the actual trajectories of the goods. As railroads are extended to connect China with Europe and Southeast Asia, as new airstrips are built in western parts of the country, and as parts of Tibet are built up in the name of national integration and stability, the geographies of former trade networks are being reshuffled. Yet traders' stories about the dirtiness of the old wool trade, the rapid production of new machine-made *pang gdan* in Sichuan, and the revitalizing of "dead" Tibetan wool also contribute to the social transformation of regional and national geographies.

In an era often characterized by the intensifying mobility of commodities and credit, the origins of some goods are seemingly becoming more and more opaque and difficult to trace. Looking at the fixing or refixing of place to objects helps us better understand the contributions of social experiences to the processes of globalization. Attaching a Tibet of a specific time and place to an object is therefore less about the knowledge of the commodity's actual origin than it is about creating and maintaining a structure of feeling amid tremendous social and economic change—sometimes for a profit. As new networks are created or older trading links are reopened to facilitate the movement of commodities, different places are valued for their economic or social capital, and new configurations of place-based narratives emerge. These different valuation systems and narratives come into conflict with the emergence of new geographical paths of capital: for instance, the silky new "urban" aprons signify a modern or sophisticated wearer; new machine-made aprons mean a loss of old virtues, a traditional way of life, and high-quality production skills.

The focus of this story now zooms in to examine how different places come to be valued in different ways—in this case, places along a portion of the Lhasa–Kalimpong trade route. The next chapter, "Silk Roads and Wool Routes," examines how traders are involved in the creation of trading places, both through the politics of naming the reopened trade route between Tibet and India and through their lived experiences of the route. It is a drama over a controversy between two groups of community members in Kalimpong: those who are for and those who are against the reopening of the Jelep-la mountain pass in addition to recently reopened Nathu-la.

CHAPTER THREE

Silk Roads and Wool Routes

What the map cuts up, the story cuts across.
Michel de Certeau, *The Practice of Everyday Life*

Tsering's Map

It is a clear weekend afternoon in Kalimpong, a town of about 43,000 mostly Nepali-speaking inhabitants in the mountainous, northernmost tip of West Bengal.[1] The rhododendrons are glaringly bright against the green foliage and the Himalayas hover on the horizon; it is especially pleasant after months of heavy monsoon rain, moldy clothes, and the sporadic landslides that prevented the delivery of provisions to the local shops for weeks on end. I am on my way to meet a man named Tsering, someone whom my friend Prakash insists I must speak with if I am doing research about the Lhasa–Kalimpong trade route.

I walk quickly down the hill, past the Mela (festival) ground and the motor stand in the center of town, where shared jeeps are available to other towns in the hill region: to Gangtok, to Kurseong, to Siliguri, to Darjeeling. I decide to pass through the haat bazaar, the twice-weekly market, on my way to meet Tsering. Most stalls feature produce, mostly from local farms—vegetables, potatoes, tofu, spices, spinach, ginger, brinjals (eggplants), tomatoes, and oranges. Then there are the bag stalls. These are piled high with bags from China, bags featuring characters that look more or less like Snoopy but with "PNOOSY" written underneath, athletic bags, and school bags with plastic zippers already in need of repair. There are rows of knives, piles of brown sugar, yeast, cheese, churpee (hard yak cheese), spices, and ready-made clothing, the latter all from China. Small household items are carefully laid out and displayed: things like small locks, batteries, nail files, plastic jewelry, bangles dusty with glitter, and posters featuring Jesus, cherubic white babies in frilly dresses, and tropical scenes with inspirational sayings like "You are so special" and "A man who spends too much is a spendthrift." My field notes soon resemble the lists of commodities that fascinated early travelers in historical accounts about Himalayan markets as well

FIGURE 17. The *Mela* (festival) ground in Kalimpong, 2006. Tina Harris.

as the oral narratives of elderly traders, except that several of the goods now come from significantly different geographical origins or directions of trade.

Tsering is a man in his early sixties who comes from a long-standing, elite Sikkimese family that was involved in the Lhasa trade in the 1940s and 1950s. Currently active in the hotel and restaurant industry in Kalimpong, he is part of a loosely defined group of merchants, hoteliers, travel agency representatives, and local officials who are lobbying the Indian government to reopen the Jelep-la mountain pass as a supplement to Nathu-la, the Sino-Indian pass only five kilometers north of Jelep-la, which was reopened for trade in June 2006.[2] The group argues that the trade route passing through Nathu-la heads straight to Gangtok, the capital of Sikkim, bypassing the town of Kalimpong, which was one of the major hubs of Tibetan trade at least until the 1960s. As Jelep-la was the route that was historically utilized for regular trade, reopening it, would, in their opinion, provide an impetus to economic activity so that "Kalimpong can regain its lost glory."

When I am first shown into Tsering's hotel office, I see him busily poring over three different maps while trying to draw the Lhasa–Kalimpong trade route on a separate blank sheet of paper, marking the location of both Nathu-la and Jelep-la. I ask him what kind of map he is making. He points to one of the official-looking maps, saying that it is "only a Sikkim map" and that it doesn't show the rest of North Bengal. He then shows me another map, which depicts all of West Bengal, but only up to the border with Sikkim, and no mountain passes appear to be labeled. Neither of these maps seems suitable for his purpose—namely, showing "the Government of India in Delhi the comparison between both routes, so they can see for themselves which one will be better." "There are no good maps!" he sighs.

Some Context: States, Borderlands, Constructs

The relationship between India and China has been featured regularly in the global media throughout the beginning of the twenty-first century, with much of the Western-centric media exuding a whiff of simultaneous fear and awe. Data such as income statistics, GDP, strategic spending, and population percentages are used to compare the rising economies of the two most populous nation-states in Asia. In a joint statement issued by Chinese premier Wen Jiabao and Indian prime minister Manmohan Singh during Wen's visit to India in 2005, bilateral India-China relations were described as a "strategic and cooperative partnership for peace and prosperity," the subtext being that such collaboration would also provide thousands of new economic opportunities for Chinese markets and Indian investors (*Economic and Political Weekly* 2005). These kinds of discussions, while prominent from the standpoint of nation-states and international relations, often create a kind of diversion whereby India and China—with their political centers in Beijing and Delhi—appear to contain their own exclusive economies and histories (van Schendel 2005). Such completely state-centered perspectives ("the Indian economy" or "the Chinese market") obscure the diverse social and economic realities of those living in the so-called margins of India and China.

Anthropologists Michiel Baud and Willem van Schendel have suggested that although "there is an extensive literature on how states have dealt with their borderlands," there is not a very developed body of work on "how borderlands have dealt with their states" (Baud and van Schendel 1997: 235). Cultural, linguistic, and trade networks originating in areas on the political edges of China and India such as Tibet, Ladakh, Arunachal Pradesh, Yunnan, Sikkim, and North Bengal have crosscut the region long before cartographers and various political powers drew up any boundary lines on a map. One approach that may be used to challenge or reposition geographical representations that derive from state-level discourses is in-depth ethnographic research with people who experience transformations in social and economic life in areas considered at the state level as "frontiers," "borderlands," or "margins." Although tracing the processes and effects of social and economic change is a difficult task, pinpointing how spatial understandings of the region are produced on various levels—from the nation-state to individuals living on the margins of the state, for example—may provide a closer understanding of how Tibet, China, and India relate to each other as spatial constructs.

This chapter examines several discourses over the reopening of the trade route that historically connected Lhasa with Kalimpong via the Jelep-la pass, stressing the importance of looking at transnational mobility across border regions and arguing that a closer examination of how various groups "fix" places is vital to the understanding of how economic change in Tibet, India, and China is actually experienced. To illustrate how new spatial representations of the region are created and anchored in geographical imaginings of the past, I follow with three brief narratives: First, I look at contending spatial and temporal discourses over the naming of the reopened Nathu-la trade route. Second, I discuss how two groups—one for and one against the reopening of Jelep-la—both agree with and struggle against state-centered conceptions of space. Finally, I examine how concerns regarding the marginalization of a place like Kalimpong (as a result of the reopening of Nathu-la) are provoking local businesspeople to make the town more visible through the marketing of local products.

Mobility and Fixity in a Globalizing World

Around the beginning of the 1990s, the drive to understand seemingly "new" features of the global economy, such as increased migration of people across borders, flexibility in labor and manufacturing practices, and the transnational movement of cultural icons such as Disney characters and Bollywood stars, fueled much excitement in the social sciences. A flurry of writing appeared in such fields as cultural studies and anthropology, detailing the intensifying interconnectedness of societies and economies across the world, where continuously mobile peoples and images were seen to constitute contemporary life, and digital technology and the Internet were hailed as providing easier and faster access to worldwide streams of information and knowledge. In particular, theories of deterritorialization (the "dissolving" or the "erasure" of political and cultural borders) and the "death" of the nation-state pictured the global cultural economy as a "complex, overlapping, disjunctive order" of flows of commodities, people, labor, money, images, and ideas (Appadurai 1991; Clifford 1997).

While many public intellectuals and academics certainly subscribe to the idea that communities and nations should not be taken for granted as bounded entities or containers of histories and cultures, a more nuanced view of global mobility and territory has emerged, influenced by the work of historians of global economies and Marxian geographers (as outlined in this book's introduction). These views, based on the premise that processes of globalization are inherently contradictory and unequal—for instance, where some people, objects, and ideas may move unfettered, others can become hindered as a result—challenge concepts such as deterritorialization, which privilege abstract and unobstructed motion across boundaries. In addition, case studies that feature ethnographic research on transnational social groups such as cross-border traders and migrants have helped demonstrate that while mobility is integral to the social and economic lives of those who are part of cross-border networks, borders themselves are by no means obsolete, and in fact state presence is felt very acutely by these groups in certain situations (again, see Das and Poole 2004; other examples of these studies are outlined in the introduction).

Again, I believe that to study the history and processes of uneven geographical mobility, it is useful to investigate how it is that different (and often competing) representations of space are produced by various stakeholders who are caught up in the rapidly changing economy of the region. More specifically, we must not privilege mobility but instead pay closer attention to the ways in which spatial mobility and fixity go hand in hand. Mobility and flows across borders—whether of finance capital, information, people, or commodities—are always "premised upon various forms of spatial fixity and localization" (Abraham and van Schendel 2005: 58). Thus, obstacles to trade may also provide opportunities for more high-risk profits, such as through the transport of illicit or restricted goods. New roads and railways, while allowing for the easier and faster movement of goods and people on one level, may simultaneously obstruct, separate, or erase other existing places; the dividing of nomadic pastures, abandonment of other passageways, or relocation of communities are examples of this. Competition over space (not simply in its physical form but also in how it is represented and remembered) results in diverse kinds of "mappings" of the region. For instance, traders may talk about their routes with reference to cross-border regions that do not coincide with a nation-based perspective. In other circumstances, however, they will at times use rhetoric that is in accord with that of the "national economies" of India and China and collaborate with states if it seems advantageous. Looking at how and why these mappings—both concrete and abstract—are created may help us understand how Tibet, China, and India (as well as other geographical spaces that may not exist on official maps) figure into the realities of individuals living along the trade route between Tibet and India, as these are regional mappings that do not always overlap with larger and more familiar urban or cosmopolitan citizenship networks (Ong 1999; Sassen 2002).

Although transnational trading groups are often characterized by anthropologists as tight-knit communities composed of a single ethnic minority who were and are somehow located in the margins of nations and therefore set apart from state-based discourses, the reality is more complex than this. Some of these groups may in fact find themselves colluding with "the majority" during various struggles over the representation of space. By paying attention to the ways in which these groups divert, reassess, and redefine their positions vis-à-vis more powerful entities, we can begin to move beyond simple binaries of "local" versus "global" and "majority" versus "minority" (Katz 1996; Lionnet and Shih 2005). To disentangle how different spatial fixing strategies represent territory on different scales, for different purposes, I now turn to two stories concerning the recent reopening of the trade route through Nathu-la.

Of Silk Roads and Wool Routes

On July 6, 2006, after forty-four years of closure, Nathu-la was officially re-opened on a damp and cold day; the ceremony was well attended by Chinese and Indian officials and media representatives.[3] The year 2006, declared by both governments as "India-China Friendship Year," marked two major infrastructural events: the inauguration of the Qinghai–Tibet railroad linking Lhasa with Beijing and the reopening of the Nathu-la border crossing between Sikkim and Tibet. From the standpoint of nation-states, these two development schemes are seen as drawing China and India closer together, both geographically and economically. In 2011, the Chinese government began extending the railroad to Shigatse in the hopes that it would be completed by 2015, and there are even rumors that it will stretch as far as Yatung, near the Indian border. On the Indian side, money was set aside to widen the Gangtok–Nathu-la road by 2012 (though road construction in Sikkim was severely hampered by the September 2011 earthquake). In 2006, representatives of both countries talked about opening the full route to tourists by 2012 or 2013. Although the majority of recent trade between India and eastern China has been conducted over truck routes through Tibet and Nepal or via maritime routes between the ports of Tianjin and Calcutta, Nathu-la offers a quick route for Chinese commodities to the port of Kolkata (only twelve hundred kilometers away), cutting transport times by more than half. Not surprisingly, this combination of accelerating economic development in China and India and new infrastructural changes has led to the creation of new spatial representations of the regional Himalayan landscape. I illustrate one of these scenarios here.

It is difficult to read a newspaper article about the reopening of Nathu-la or to visit North Bengal or Sikkim without noticing multiple references to the revival of the "ancient Silk Road." On a recent trip to the region, for example, I glimpsed several large billboards on the side of Highway 31A (the winding road that leads to the center of Gangtok) sponsored by the Department of Commerce and Industries of the Government of Sikkim to promote the reopening of Nathu-la. One read: "Once the Silk Route . . . Now the Road to a Golden Future." Another simply said, "Indo-China Border Trade." In the months building up to the reopening of the pass and immediately thereafter, articles in the Chinese, Indian, and international presses featured headlines such as "Silk Road Revived as Border Pass Reopens After 44 Years" and "History Made as Silk Road Is Reopened for Border Trade" (China Daily 2006; Hussain 2006; India E-News 2006). While conversing with a local official in Sikkim, I noticed that he often referred to the route between Lhasa and Gangtok as the "ancient" or "traditional" Silk Road. Although most people did not use this phrase, some traders in both Tibet and India shared the opinion that the comparisons between the Nathu-la route and the Silk Road were far-fetched. For instance, one businessman from a former trading family in Kalimpong responded to the media fanfare by stating that the Silk Road was located far away, nowhere near the route through Nathu-la: "But this was never the Silk Route!" he exclaimed. "If you are going to call it something, you should call it the Wool Route." He stressed the fact that wool—not silk—was the most significant commodity to be transported down the roads from Tibet to India, and that highlighting this admittedly duller-sounding commodity would have been a much more accurate representation of the legacy of trade in the region.

I use this account of the *naming* of the trade route during the reopening of Nathu-la as one example of the competing discourses over the production of spatial and temporal representations of the rapidly transforming region. The term "Silk Roads" (*Seidenstraßen*) is said to have been first introduced by a German geographer, Ferdinand von Richthofen, in the nineteenth century (Christian 2000). It has been used to describe the network of routes connecting northern China and Central Asia with the Mediterranean, particularly during

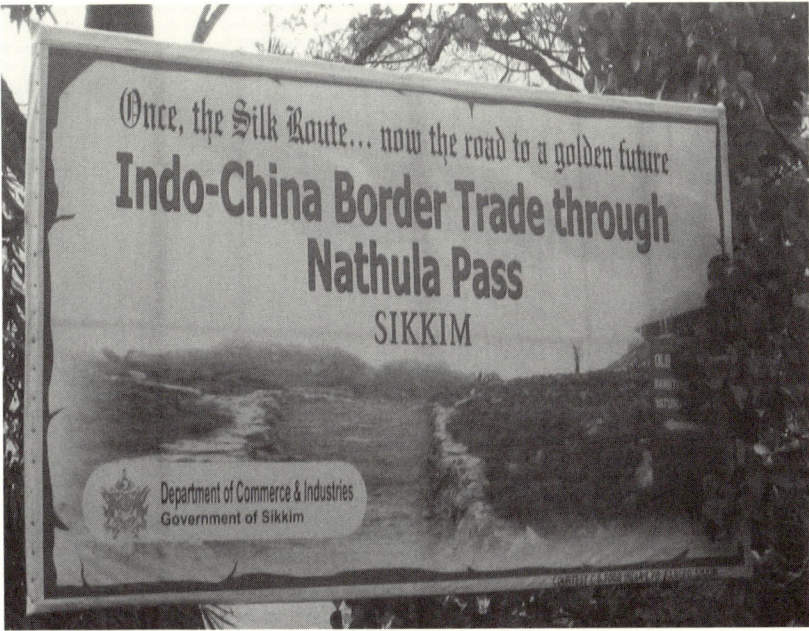

FIGURE 18. Sign on National Highway 31A, on the road to Gangtok. Mark Turin.

FIGURE 19. Sign at Nathu-la, 2006. Pema Wangchuk.

the Tang Dynasty in the seventh through tenth centuries, and it often evokes the popular imagery of camel caravans traversing the Gobi Desert, laden with valuable silks, perfumes, and porcelain (Christian 2000). More a web of separate tributaries controlled by numerous middlemen than a single road, the trade included maritime routes that stretched as far as Japan and overland paths that extended over the Himalayas and south to India. Silk was but one of the numerous (luxury) commodities to be carried along the routes; domesticated animals, gunpowder, paper, and medicinal herbs were also exchanged, alongside the dissemination of religious, artistic, and musical knowledge. Yet the official excitement over reopening Nathu-la as part of the "Silk Road" is not so much a revival of ancient connections as it is part of a *contemporary* global discourse. It comes at a time when regions in Asia previously thought of as "remote" are becoming part of major global infrastructural and information networks as well as intra-Asian and international tourist destinations. Asian and other investors in the construction of new oil pipelines across Central Asia, the inauguration of the Qinghai–Tibet railroad, and the encouragement of tourism capitalize on the imagery of the ancient Silk Road networks as a metaphor for a "return" to an Asia-centered globalization.

On one hand, the use of the term *Silk Road* to describe the trade route from Tibet to India through Nathu-la extends one's idea of the territory, making the small Himalayan region part of a larger global narrative of Silk Road revitalization, of mapping "uncharted" terrains and connecting the region to urban China's rise in the global economy. At the same time, the use of the term *Wool Route* by the Kalimpong businessman makes visible a lesser-known, smaller-scale Himalayan geography characterized by the trade in wool. Although silk was in fact one of the many commodities that was transported from China via Tibet to India prior to the closing of Jelep-la, it was by no means as central to the economy of the region as wool—or even yak tails. Wool undeniably came through the route from regions like Yamdrok or Changtang in Tibet; silk was (and is) associated with Sichuan Province in inland western China. Since commodities are often associated with their place of origin (whether or not they actually come from there), the naming of a specific product like silk or wool becomes another marker of place, pointing to divergent and competing social, economic, and political histories.

These two examples of place-making, one based on silk, the other on wool, are, at the same time, temporally significant. During the opening ceremony of Nathu-la, Pawan Chamling, the Chief Minister of Sikkim, began his speech with the following statement: "A contact that started centuries back between our two civilizations is being reestablished today" (Rajesh 2006). Here, the Silk Road narrative evokes Chinese and Indian prehistories, maintaining that their connection is an ancient one. And yet, even though the nation-state actually came into being as a concept as recently as the 1950s, "civilizations" are spoken of as if they had always been state-centered (Ludden 2003).

Obscured by this official state-based representation of the temporality of the Silk Road, the narrative of wool is more recent, calling to mind a period of twentieth-century regional economic history checkered by political contention between Britain and India, India and China, and of course separatist movements in Tibet, Sikkim, and West Bengal.[4] Nevertheless, it must be emphasized that these traders' narratives are not always emblematic of some sort of obvious or conscious "resistance" against state-centered discourses, for traders also capitalize on the Silk Road imagery by labeling their shops with such straightforward names as "The Silk Road Trading Company." But in fact, many people—in spite of Indian restrictions on imports of Chinese goods through Nathu-la— would rather trade more marketable high-profit goods like Chinese-made electronics and kitchen appliances than the officially sanctioned wool and yak tails. The list of restricted commodities for imports to India includes goatskin, wool, yak tails, silk, butter, and salt. Exports include cloth, blankets, medicinal herbs, bicycles, rice, and tea. All told, traders map out an economic history of the region that at times diverges from that of the state, and at times overlaps. It is important to pinpoint where and when these alternative mappings are produced in order to better understand the cultural and economic realities of the region are experienced. The next story will exemplify this further.

"Delhi Doesn't Understand": Jelep-la versus Nathu-la

If on the national and global level, the reopening of Nathu-la reaffirms "ancient" linkages between China and India and other parts of the world, on other levels, it dislocates smaller-scale social and economic networks. In contrast to the pre-1962 days when mule caravans followed the trade route that led from Lhasa to Kalimpong via Jelep-la, today's renewed trade circuit through Nathu-la links Lhasa more immediately to Gangtok. Now, trucks that bring Chinese commodities through Gangtok down Highway 31A bypass formerly important trading towns like Kalimpong, eventually reaching Siliguri in the plains, the current major hub in West Bengal for goods heading to and from the Northeast Indian states, Bhutan, Nepal, and farther south to the port city of Kolkata.

Whether or not the volume of trade through Nathu-la ever becomes "success-ful" (it has gotten off to a very slow start—according to the Yadong County Chamber of Commerce, only 1,217 Chinese traders entered the Indian side and 574 Indian traders entered the Chinese side during the first trading season in 2006), this geographical transformation has altered the ways in which various factions explain the *future* economic prospects of the landscape between India and Tibet.

After the trade passes were closed in 1962, Kalimpong never quite "made it," notwithstanding attempts to reinvent itself as a high-profile tourist town or a tea-producing center like its larger neighbor, Darjeeling. In 1975, after Sikkim became a separate state and as Gangtok began to grow in size and resources, Ka-limpong—although still a relatively wealthy middle-class town—experienced additional economic fallout as the tea industry began to fall into decline, and serious separatist and police clashes in the 1980s kept away would-be holiday-goers. During the late 1990s and early years of the twenty-first century, local crops such as cardamom and oranges suffered low yields due to a virus. Also, as a few residents noted, a high percentage of the English-medium educated youth left Kalimpong for employment opportunities in call centers in urban areas like Bangalore and Kolkata, returning to the hills only to visit their relatives for brief holidays. Several residents linked an increase in theft, drug use, and sexual as-sault to the problem of bored, unemployed, mostly male youth who were "left behind," either unable or unwilling to find jobs in larger cities.

In 2006, Tsering's group of merchants, traders, hotel owners, and local offi-cials in the Kalimpong area began to meet together informally in order to lobby local and state officials to reopen Jelep-la in addition to Nathu-la. They argued that during the days of the wool trade, Jelep-la was a much more significant pass than Nathu-la and brought in a higher volume of trade. Furthermore, un-like Nathu-la, which can be accessed only during warmer months, Jelep-la is an all-weather, all-season pass. It is often translated into English as "easy pass." Opening Jelep-la, I was told, would help Kalimpong revive the "Golden Era" of trade, bringing in much-needed revenue for a flagging economy, and would provide opportunities to market Kalimpong as a Himalayan tourist destina-tion. Soon after the pro–Jelep-la group formed, however, a rival group emerged, claiming that increased trade would in fact cause considerable environmental damage to the already fragile hill topography.

Tsering, as one member of the group pushing for the opening of Jelep-la, mentioned that the number of tourists at his hotel had indeed declined sig-nificantly over the previous ten to twenty years. Frustrated with the growing economic and social decline of the region, including the theft and drug use, he joined the collective of Kalimpong residents, local officials, and businesspeople who were promoting the reopening of the Jelep-la route. While excitedly describing a proposal the group planned to send to officials in Delhi, he remarked that "trade through Jelep-la would be the biggest thing that could happen to this region!" At first glance, Tsering's push to reopen Jelep-la coincides with mainstream development discourses. He spoke at length about how development maintains stability in a politically volatile region and said that the reopened pass would bring in tourism, commodities, and more jobs for the unemployed. And yet when I asked how the town would be able to sustain the construction of roads and the increase in traffic along the mountain highways during the monsoon season, he replied: "Opening up Jelep-la is a desperate move so that we can exist economically. I'm not talking about social or spiritual development. We'll lose that. When Nathu-la opened in July, the tribal people there [in Sikkim] said the same thing, that it would destroy the environment. It [destruction] is inevitable . . . I am a Communist emotionally, but I need to be a capitalist in order to make a living."[5]

Another perspective comes from a member of the group fighting against the proposed reopening of Jelep-la. Ramesh is a middle-aged Nepali man, a long-time resident of Kalimpong who currently works at a small local school. He was extremely concerned by the hopeful air earlier given off by traders and residents of Gangtok regarding the reopening of Nathu-la:

> How could they be hopeful? It is just not practical. I witnessed the floods of 1967 . . . We will see such environmental degradation. I read an article which talked about the railway to Lhasa and how it will open up opportunities here. It's easy for some guy in Delhi to sit there in his office imagining that, but no way can all those trucks come through here. It's not practical. The environmental destruction will be immense. Remember when the road was destroyed in the monsoon [in September 2006]? I left Siliguri at 5:00 p.m. and got back to Kalimpong at 1:30 a.m. [normally only a two-and-a-half-hour journey]. I don't see how this can happen. And you've seen the Teesta dam project, we don't need any more of that.[6] I am one of those against the trade.

What seemed at first to be a simple set of opposing positions—one group in favor of the reopening of Jelep-la because it would be economically beneficial to Kalimpong and the region, and one group against it because it would be environmentally detrimental—turned out to be more complex, and in fact, both groups shared very similar frustrations that were manifested geographically. They felt that Kalimpong was "losing out" because Gangtok would be the immediate beneficiary of trade through Nathu-la with China, and that the entire hill region's economic and ecological conditions were being ignored by larger state apparatuses—those in West Bengal and those in Delhi. Although the group pushing for the reopening of Jelep-la paralleled development-driven sentiments by claiming that "trade brings prosperity," proponents often noted in the same breath that people in Kalimpong felt left out of larger political decisions, and that any proposed development in the region needed to be conducted with a keen awareness of the specific environmental concerns of the town. One trader noted that in the 1990s, "Kalimpong people lobbied to get Jelep-la opened, but all we are, are a subdivision, just a small town with no power in West Bengal. Sikkim has a much stronger lobby because it's an entire state." He went on to say that policy makers and development officials who come from Delhi, "or even from the state government, or the plains," cannot suggest appropriate long-term plans because "Delhi's homework has not been done for Kalimpong" and they do not know what "life is like in the hills." As these residents have implied, the state-based policy makers and government officials have failed to understand the economic and geographic realities of the region, an area that is not demarcated by lines on an official map. "Delhi just doesn't understand" was a sentiment I often heard from both groups.

As these narratives demonstrate, landscape is always in a state of becoming and is never entirely stable (D. Mitchell 1996). Territory can shift, expand, and contract based on various groups' needs. To open new markets for Indian and Chinese goods, for instance, and to cut the time for their transport in half, the mountain range between Tibet and India had to be overcome for capital to continue to move. But this very notion of "opening" or "reopening" passes through the mountainous border—accompanied by the raising of national flags, the symbolic handshakes of representatives from the two states, and the erection of gates and passport checkpoints—paradoxically reinforces the Indian and Chinese states as stable, bounded entities. If the conception of the state is, as Timothy Mitchell and others have claimed, more of a "structural effect" that is "reproduced in visible everyday forms, such as the language of legal practice . . . or the marking and policing of frontiers" than an actual "apparatus that stands apart from the rest of the social world," then the production of opposing, alternative spatial articulations can be constitutive of power as well (T. Mitchell 1999: 173, 180; see also Abrams 1988 and Foucault 1979). In this brief example of the Wool Route versus the Silk Road, trading groups make visible or "fix" alternative spaces against other, more authoritative and controlling representations of space, pointing to lived histories based along a Tibetan trade route that crosscuts and, to some extent, defies state-produced boundaries.

A Geographical Excuse

When Tsering exclaimed that there were "no good maps," it was because he had to make his case to Delhi through official maps. The lived geographical under-standing of the economic realities of the region between India and China—not found in any "official" discourse—can be used strategically in some situations to challenge more abstract, state-centered representations of the route that serve to transform the landscape into a conduit for the "free" flow of capital. Such a challenge is reminiscent of what K. C., a schoolteacher in Kalimpong, once said about "geographical excuses."

K. C. and I were in a small hotel restaurant, eating steamed dumplings made with cheese from a local dairy and discussing the history of products and ser-vices—flowers, dairy, fruit, tea, tourism, and timber—that Kalimpong had more recently been famous for, in addition to the wool trade. K. C. mentioned that up until the late 1950s and early 1960s, the town of Siliguri—currently the largest and busiest hub in the North Bengal plains for rail, truck, and air trans-port between India, Nepal, Bangladesh, and Bhutan—was a tiny place, with "no university, not even any medical facilities."

Prior to that time, according to K. C., economic life in northern Bengal had been centered on the trade in the hills, in Kalimpong and Darjeeling. Two fac-tors, he said, changed the balance of the trading landscape: "One, the 1962 bor-der closing, and two, political issues."[7] Now, he laments, Siliguri has "taken over as the main hub of trade. Everyone makes the excuse that Siliguri is perfect because of its geographical location. So why was it not the trade hub prior to 1962? It was still in the same geographical location!" He continued to explain that if transportation in the hills along the Lhasa–Kalimpong route had been such an issue, there would have been no trade boom in the early twentieth cen-tury. People were able to traverse the mountains back then with no problem. In fact, according to K. C., the West Bengal state has now "become indifferent to the hills." The government says, "'the plains are better, geographically,' but there's not much difference. *This is just a geographical excuse*" (my emphasis).

Saying that Siliguri's location as a contemporary trading hub is a mere "geo-graphical excuse" reflects his concerns that state control over the movement and direction of trade in the region will continue to shift the balance lopsidedly toward the plains, despite the hill towns having been extremely prosperous in comparison with the plains during the peak of the wool trade in the early and mid-twentieth century. The historical balance of economic power in the re-gion, while at first concentrated in the British hill towns like Kalimpong and then replaced by upper- and middle-class Indians with political and economic ties to the former colonial elite, began to shift after the 1978 reforms in China. Now, the reopened route from China through the Nathu-la mountain pass reinforces this latest trajectory, with Siliguri reaping a fair amount of the benefits of the new Chinese trade. The route wends its way through Gangtok, bypassing Kalimpong and several other northern Bengal hill towns as it heads down to Siliguri, a direct route to the port of Kolkata.

K. C. went on to say that products such as tea and timber, once important to the economy of the hill towns like Kalimpong and Darjeeling, are now "plucked from this region" and transported straight to Siliguri:

> If people in the hills need these items, they are purchased in Siliguri and then transported back again to the hills! The infrastructure is not in Kalimpong anymore, not in the hills anymore. It is in Siliguri—all the warehouses for building materials. There are no big auction houses here. Siliguri has become the main distribution center. Now there is no one in the state government representing the hill areas of Darjeeling or Kalimpong. The representative is from Siliguri, the plains, and will only think about how trade will affect Siliguri, not Kalimpong.

As K. C. talked about the formerly important trade items of the region's boom times—tea, timber, and tourism, sometimes called the "three *t*s of the hills"—he became more pessimistic. "There's no future in tea, timber, and tourism. We need to think of [promoting] other products, local products like ginger and oranges." He spent the rest of the conversation animatedly discussing the various ways the state might assist local individuals to harvest fruits or run dairies:

> [Ginger] would be a great product to sell over Nathu-la if the state just paid attention and helped us cure the infection.[8] Similarly, the region has some lovely oranges. Very sweet oranges, but they were also affected by a virus in the last five years. They used to come from villages. Also there used to be an orange-juice-producing plant which has since closed down. See the apples in the market? They come all the way from Kashmir, and some from Bhutan. If we could do the same with oranges, then we could gain a foothold in the market. The Swiss dairy used to be here, but [it] closed because of a bureaucratic hassle or something. One of the good things that has come out of this is that it has forced the local people to do something about it. Some local people have taken up small private dairies. Cheese, milk, lollipops; the negative factor has pushed people to do this. There are about twenty to twenty-five dairies here—well, as opposed to forty to sixty back in the heyday. Hind's and Lark's [names of local shops] sell the homemade cheese. Cheese is sold to other towns like Darjeeling. Also, plums and peaches

would be nice; the land has great potential for producing plums and peaches as potential cash crops. The thing is, the state is just concentrating on security and agitators in this region and is not focused on the daily life of the people. The state needs to be a facilitator, not an interferer.

If the "three *ts*" are outdated (that is, not profitable anymore) for producers in Kalimpong, K. C. demonstrates that there are spaces for local farmers and cheese makers to make their town relevant again in the face of the new economy, an economy marked by increased competition between places and "place branding" for profit. Reattaching objects to place (even though they may not originally be produced in those places) and promoting another, equally viable trade route such as Jelep-la are examples of ways people can attempt to make their locations more visible or coherent in the face of the state, while at the same time engaging in competition with other local or regional economies. Rather as Pema did with his aprons, K. C. considers how to make the past economic glory of Kalimpong shine again via renewed investment in local foodstuffs vis-à-vis the new state-directed routes. This tack seems to reflect a desire to create a trading diversion that promotes and encourages new kinds of local or regional spatial fixes against the existing state-sanctioned ones. Thus, in order to explore more carefully the complexities of "making visible" or fixing certain places vis-à-vis the state, the next chapter focuses on ethnographic examples of traders' border-crossing experiences, showing *how* people actually fix places and how both the discourses of state security and individual experiences of security are highlighted across the border zones of China, India, and Nepal.

CHAPTER FOUR

Reopenings and Restrictions

We are good friends / Women shi hao pengyou
> *One of several phrases on a sign encouraging tourists to learn*
> *Chinese phrases on the Indian side of the Nathu-la border*

Border Crossings

I begin this chapter with two scenes that take place at border crossings.

1. *Nathu-la.* For decades, the hills of North Bengal and Sikkim have provided a cool escape for many middle-class tourists (and formerly, British colonists) during the stifling Indian summers. Although currently restricted to Indian citizens with special one-day permits, one common scenic destination is the India-China border post at Nathu-la. Travel agencies in Kalimpong, Darjeeling, and Gangtok advertise the Nathu-la tour package as a high-altitude adventure; according to one Indian travel agency, it is a "wonderful place to behold the nature's splendor and admire the armed forces that stand without the fear of sun or rain to guard their country" (Bharat Online). Indeed, any search for "Nathu-la" on the YouTube video-sharing website will bring up multiple home video clips of groups of Bengali tourists visiting the border. After watching a few of these videos, I noticed strikingly similar footage in many of the clips.

Permit me to merge them into an archetypal scene: At the border post at thirteen thousand feet, two small children in brown wool balaclavas gingerly poke their feet though the barbed-wire fence as their father films their bodies in India and their toes in China. Then, through the ghostly high-altitude mist and fog, the camera pans slowly across the bleak landscape. In a small sentry box in the background, a Chinese guard is dimly visible. The camera continues to pan across the pass and soon we see families on the Indian side standing in line to take photographs and shake hands with the Chinese soldiers on the other side of the wire fence.

2. *The Friendship Bridge.* Hundreds of miles west of the Nathu-la border post is the Sino-Nepalese Friendship Bridge and its two corresponding passport and

customs checkpoints in Tibet and Nepal. For both tourists and traders, this is the most frequented overland route from Lhasa to Kathmandu. Every morning at 8:00 a.m., jeeps line up in Dram/Zhangmu/Khasa (Tibetan, Chinese, and Nepali names for the same town) on the Tibet side of the bridge and in Kodari on the Nepal side, waiting to transport travelers on their journeys across the border. In Dram, I stand in front of the red and white barber pole–like border barrier with a German tour group and a number of western trekkers heading home from their trip to Everest base camp, waiting for the border office to open so that our passports can be marked with a China exit stamp. We wait in line for one hour, watching local women traders duck under the barrier in full view of the Chinese guards and walk across the bridge carrying baskets filled with vegetables. Once we get our passports stamped and begin to walk across the bridge to Nepal, some tourists stop to take photographs with their legs splayed on either side of the red line in the middle of the bridge. One side says "China," the other, "Nepal."

This chapter examines practices of border-crossing along the Lhasa–Kathmandu–Kalimpong trade route. Although the kinds of practices undertaken by tourists (such as passport checks and photographs with Chinese guards) and those of the women traders (for example, slipping under the barrier to trade vegetables) are both examples of the everyday experiences of crossing national boundaries, they reveal differential relationships with the mechanisms that maintain state power. Barriers are fixed to mark and securitize national territory, yet these state-based fixing processes often go hand-in-hand with the notion of economic prosperity based on "free flows" of global trade. There are frequent tensions between state-based border restrictions and the actual movements of people across the border. Through the material practices of border-crossers acting with and around these restrictions, alternative trade routes are produced that do not necessarily overlap with state ideas of what the routes should look like. Ultimately, the goal of this chapter is to show that borders, like the routes discussed in chapter 3, are represented by the state in one way, but actually *lived* in others. Even in the two opening descriptions, the tensions and contradictions between lived experience and state power are evident; for example, the tourists' fear of stepping too far into forbidden territory, the border crossing poignantly named the "Friendship" Bridge, and the local women who duck under the border barriers daily to continue trade.

I begin with a brief outline of some of the major claims made in the field of border studies, as well as some departures I take from this literature. I then turn to specific narratives of trade practices across Chinese, Nepali, and Indian borders and attempt to demonstrate how these very practices create trading places and routes that are integral to the history of the region. The trading practices described are divided into three sections: Reconnecting, (Il)licit Commodities, and Identification. Toward the end of the chapter, I revert to a regional scale, showing how the practices just described are manifested geographically and reflect increasing economic and social gaps between "small" and "big" cross-border traders.

Borderland Economies

Perhaps with the exception of "Tibet, the Roof of the World," most representations of the region I am concerned with are rather unflattering.[1] Take for instance Sikkim and North Bengal, which make up a strip in the northeast of India bordered by Nepal, Tibet, Bangladesh, and Bhutan, sometimes called the "chicken's neck" of India, or, in a tongue-in-cheek article by Kanak Dixit of *Himal Southasian*, "the mouth of the tube of toothpaste that is India, squeezed out through this orifice, to exude into the eight states of the Indian Northeast" (Dixit 2002). Further, "Nepal likes to think of itself as the proverbial yam sandwiched between two boulders" (that is, China and India) (Turin and Shneiderman 2003: 8). The attachment of names or particular images to places is laden with meaning; in these cases, certain areas are depicted as marginal, or at least as not important to the surrounding territory. According to Rob Shields (1991: 3), "marginal places are not necessarily on geographical peripheries, but they have been placed on the periphery of cultural systems of space." Later in this chapter, I look at how people use the status of geographical marginality to achieve economic goals.

Those who work in Tibet and among Tibetan-speaking communities often find themselves having to justify their existence within university or other institutional area-studies departments. The Tibetan-speaking Himalayan region—which includes Tibet (the Tibet Autonomous Region, or the TAR, currently under Chinese administration) plus what some scholars call "historical" or "cultural" Tibet (that is, the Chinese provinces of Gansu, Sichuan, Qinghai, and Yunnan), as well as parts of Bhutan, northern Nepal, northern and northeastern India, and Burma—does not conveniently fit into currently familiar area-studies categories such as "East Asia," "Central Asia," "Southeast Asia," and "South Asia." It is precisely this area's lack of conformity to existing conventional geographical configurations (whether as a nation-state or as part of "South Asia" or "East Asia") that creates the need for a more nuanced exploration of its important social and economic linkages. The question "Why doesn't this region qualify as an institutional 'area'?" has been outlined by Willem van Schendel in an article on rethinking area studies in Asia. Van Schendel presents an interesting heuristic idea: he asks his readers to consider an area of which no map can be found; this area, which he dubs "Zomia," includes patches of Burma, northeastern India, Bhutan, China, Nepal, Tibet, and Bangladesh (van Schendel 2002: 647–68). Representing an alternative spatial scale that is neither national nor global, he shows that this regional scale matters significantly for many people living in the area because of various long-standing economic, social, and cultural connections (Shneiderman 2010; van Schendel 2002). Following others who have maintained that frontiers are shaped less by geographical conditions than by the impact of those who created them, we need to look at transformations of territory in spaces that national maps conceal or ignore (Lattimore 1962: 384; Ludden 2003: 1070). If spaces are not simply "there" but are instead produced by those who have various claims over them, the challenge is to investigate in detail how nonmainstream representations of space are rooted in history and produced, conditioned, structured, and experienced in everyday life. This is especially significant for those living in the region that includes Sikkim and northern Bengal, northeastern India, Bhutan, eastern Nepal, and Tibet, often obscured by contemporary discourses that focus on macro-level analyses of the "rising economies of India and China."

The writing on cross-border identities and capitalism has grown quite extensive since Benedict Anderson's work on long-distance nationalism (Anderson 1992). Though the bulk of research on trade or security along border areas is more or less still focused on the us-Mexico border, there are good case-based studies of social actors and the state in other border regions, such as Lynn Stephen's study of Oaxacan transborder communities and Carolyn Nordstrom's work on global smuggling across economic war zones (Nordstrom 2007; Stephen 2007).[2] And yet, there is still a need to tie together broader, geographically informed perspectives on transnational trade to ethnographic studies of cross-border experiences, such as narratives of transnational traders or the logistics of transport in border regions. Conceptualizations that emerge from anthropology *and* geography are useful for building connections between material life and the study of long-term geographical shifts of capital, directly addressing a seemingly simple question raised by scholars of globalization: how might one connect studies of small-scale (or "local") social practices with more abstract theories based on economic change at a global level? This interdisciplinary attempt to examine the interconnections between global historical change and daily interactions with material life is therefore partly a reaction to more geographically limiting subfields such as urban anthropology or area studies.

Historically, various disciplines' critiques of political economy and globalization have had a significant influence on studies of borderland trade.[3] Within this body of work, scholarship on world historical systems, such as Philip Curtin's study on world trade, Fernand Braudel's writing on European capitalism, and Owen Lattimore's work on inner Asian frontiers, features large-scale processes of place-making without losing sight of the particularities of "material life," grounding macro-historical work in the detailed analyses of everyday experiences with the exchange of commodities. Relatively neglected by contemporary social scientists (save for his role in the McCarthy indictments), Lattimore offered a historical understanding of large-scale capitalist change in inner Asia that stemmed from his early work on nomad traders in Mongolia and his personal experience as a traveling salesman of British goods in China. As Lattimore himself puts it in the preface to *Studies in Frontier History and Society*, he looked to "treat history not as a series of invasions and pressures but as an interacting mode" (Lattimore 1962: 30). I wish to follow in these footsteps, demonstrating why the use of ethnography is integral to understanding large-scale shifts in capital in a region that is still relatively unrepresented in macro histories.

Although still a relatively small field, comparative studies of borderlands in the social sciences have been useful in challenging state-centered conceptions of space. In these studies (which are necessarily cross-regional and cross-disciplinary), borders are treated as cultural zones or identified as being re-classified into economic "seams" that privilege the "priority of the market" and stretch across nations, as opposed to hard-and-fast lines of political and geographical separation (Cowen 2010). The idea of a border is seen as itself fraught with contradictions: it separates some communities while reinforcing the unity of others; it connects at the same time as it disconnects. For instance, in case studies based on the US-Mexico frontier, Michael Kearney questions why the budget for US border control consists of millions of dollars, but no money is allocated to repair the fence (Kearney 1991: 57). The answer, he says, lies in state surveillance where such activities are not for prevention but are in actuality "a way to discipline" (Kearney 1991: 61). Border paradoxes such as these come up time and time again in the literature and—as the discussion in chapter 1 of the British expansion into the Chumbi Valley in the early twentieth century illustrates—are historically not new.

Those who cross borders do not always defy the state, but often comply with it in a variety of ways in order to continue to conduct business. In an account of Moroccan border crime, McMurray portrays smugglers as not necessarily employing a kind of everyday resistance to state power even when they plead with the border guards for leniency and eventually "bluff" the guards. Instead, both the guards and the smugglers can be considered part of a border pageant in which subservience to the state is enacted publicly (McMurray 2003: 142). Similarly, with migrant laborers who cross the Malaysian-Indonesian border- land and compete for work, "not only do local people draw upon the forms and practices of the state for their own ends, but in so doing they transform what is meant by the state and state practice" (Amster 2005: 38).[4]

I follow up on this useful and significant literature, but first I wish to empha- size three lines of departure based on some minor qualms I have about existing work in this field. First, as outlined in previous chapters, I find that studies of mobility, globalization, and borders tend to "privileg[e] theories of displace- ment over location" (Behdad 2005: 231). Perhaps excepting criticisms of deterri- torialization, the literature emphasizes a sort of unlimited fluidity of people and goods across borders. The celebratory use of tropes of utopian mobility mask the very real experiences of how people and goods cross border markers and checkpoints in the first place. There is, therefore, a need to pay equal attention to the politics of location, examining the need of those in power to fix, stabi- lize, or make visible borders for the movement of capital and people and the reinforcement of the nation-state, as well as different kinds of oppositional or alternative spatial mechanisms established by those struggling to make a living in border areas. In other words, to "think relationally, to develop a dialectical understanding of the dichotomy mobility/immobility and put it into motion" (Franquesa 2011: 1019).

Second, although borderlands are often treated by scholars as exceptional or liminal places, they are not always exceptions for those people who are living in the borderlands. The danger of privileging national border zones as inher- ent states of exception (or as inherently exceptional states) is that it ignores other kinds of borders—those not necessarily limited to delineating nations and physical territory. These include, for example, divisions between class and ethnic "ghettos" within a city; or between rural and urban areas, even mani- fested through differences in clothing style, as we saw with the aprons in chap- ter 2. What may be unusual—if not exceptional—about border zones is not that they are inherently different as physical, geographical places, but since they are state-imposed and controlled, they emerge as zones of contradiction when ev- eryday practices across borders do not "fit" into what is expected or controlled by the state.

Third, and finally, I wish to emphasize that it is relatively easy for certain people to cross certain borders with certain goods at certain times. There is a great need to go beyond simply characterizing borders as zones of contradiction, tension, or paradox. Untangling some of the contradictions and unraveling the multiple tensions between economic mobility and fixity and separation and unity in border areas will, I hope, help pinpoint where differential experiences of border trading might lie. I now turn to an example from my fieldwork that illustrates what border-studies theorists might mean by "the paradoxes of the border."

At one point during my research, I found myself having tea and sweets at the home of an ex-member of the Sikkim Chamber of Commerce, who was introduced to me by a wealthy trader I had interviewed a day earlier. A major spokesperson for the promotion of Nathu-la trade, Mr. Sen was hospitable and seemed excited about and supportive of my research topic. He informed me that "the opening of Nathu-la is like when the wall between East and West Germany was demolished! . . . I have a dream, that in the long horizon, Tibetans in Tibet will have more economic opportunities, and they will slowly gain strength. Then when they are free to trade anywhere, through India, through the world, they will become independent again!"

And yet, ten minutes later, Sen remarked, "I still believe that traders shouldn't move freely, though. Because traders work for profit, not for the nation. They work as individuals, for their own profit. We in India need to assess what goods are surplus and what we are short of, then we can control trade properly. For us, the nation is first, not profit."

I draw attention to this conversation not to accuse Sen of hypocritically shifting between positions, but to point out that the contradictions in this dialogue are emblematic of state-level policies on border trade. Here, Sen's brand of neoliberal Sino-Indian free trade goes hand-in-hand with the drive for Tibetan sovereignty as well as strict control over the traders' activities. And still, Sen's ideas stand in contrast to the official Chinese line on Tibet, which is that economic development in Tibet will increase the stability of the PRC as a whole. According to Chinese officials, the Great Western Development program is "the necessary choice for solving China's nationality problems under the new historical conditions . . . [E]ach nationality will become more strongly unified with each day under the centripetal force . . . of the large family of the Chinese nation" (Li 2000).

Peter Andreas discusses these contradictions in relation to what he calls the public "performance" of US-Mexico border control. Andreas notes that states must "assure some of the audience that the border is being opened to legal flows while reassuring the rest of the audience that it is being sufficiently closed to illegal flows" (Andreas 2000: 10, 17). It is not "abstract market forces" that are at work here, but diversions—geographical, economic, social—produced by the state and encountered by traders at the border (Andreas 2000). Similarly, in a study of trade in the Golden Triangle region of Southeast Asia, Andrew Walker argues that "as trading becomes more liberalized, opportunities for regulation flourish" (Walker 1999: 7). Thus, the very act of "opening" or "reopening" a border for "free trade" paradoxically allows for regulatory and disciplinary mechanisms that reinforce the image of the Indian and Chinese states as bounded entities.

For instance, after Nathu-la was reopened for trade on July 6, 2006, those on the Indian side found that the wire fences (first erected after the border was shut in 1962) were replaced by twenty-meter stone walls. Traders had to cross seven security checkpoints from Gangtok to the border, and there were long bureaucratic delays in obtaining the necessary import-export codes, as well as strict restrictions on who could apply for trade permits.[5] In 2006, traders were allowed to sell goods worth only 25,000 Indian rupees a day (about US$500), and a fair amount of their profits ended up going toward the rental of four-wheeled vehicles that could manage the uneven terrain from Gangtok to the pass. On the Tibet side, they found they had to hire Chinese vehicles to transport them to the trade mart on the other side. Furthermore, traders had to conform to official state-imposed modes of temporality: business is allowed only from June to September, Monday through Thursday, 7:30 a.m. to 3:30 p.m. The traders interviewed also noted that the monsoon rains often delay their journeys, allowing for a paltry two hours of trade per day (Atreya 2007). At the time of this writing, in 2011, traders from the Indian side are allowed to import only fifteen kinds of commodities and export only twenty-nine. Curiously, the list of imports matches that of the pre-1962 trade: yak tails, salt, wool, butter. Many traders laugh at this list, frustrated that they cannot buy or sell more profitable goods such as mobile phones or DVD players like some of the wealthier *thulo maanchhe* (Nep., "big men" or "big traders") who directly tap into Chinese networks by flying straight to Guangzhou to do business in the hundreds of large wholesale warehouses there.

Ram, a Sikkim-based trader of Ayurvedic medicines and rice, has had a relatively successful run after the initial reopening of Nathu-la, even receiving a certificate from the PRC for "best trader." Since the Nathu-la pass is open for business only during warmer, non-monsoon months (May or June through September), some Indian traders head to Kathmandu during the off-season to continue their cross-border trade with China. However, Ram finds it a bit difficult to move so often, saying that the traders who relocate to Nepal "are more interested in business than us, us people who trade seasonally." Upon arriving at the Tibet border for the first time in 2006, Ram says that he and the other traders from Sikkim "weren't told about the rules. Both sides didn't know what kinds of items could and couldn't be sold. The IEC [Import Export Code] was mandatory for everyone, but no Sikkimese traders had this. It took India one week to resolve this. You can only trade 25,000 rupees, so trade started very slowly. This was a problem between the Central Government in Delhi and the Sikkimese state—it was all internal politics. There is no direct tax in Sikkim, but in order to get an IEC, you need the direct tax law."

Eventually, an exemption was made for Sikkimese traders to obtain the IECS required by the national government, but this story is reminiscent of the earlier refrain, "Delhi just doesn't understand." It is but one of many examples of the miscommunication that occurs in border zones between traders, officials in charge of local or regional regulatory mechanisms, and national-level bureaucrats, not to mention the confusion over what kinds of goods in what quantities can be brought across the border on each side.

Another trader in Sikkim showed me a photograph of the goods available in the trade marts on both sides of the border. The glass cases in the Chinese mart were packed full of items such as souvenir yak-tail fly whisks, yogurt drinks, shawls, and cans of Pabst Blue Ribbon beer. On top of this photo he placed another photo from the trade mart on the Indian side of the border. The shelves on the Indian side were nearly empty, with a pile of three sad-looking onions and potatoes and what looked like Sikkimese boxes of tea. He made a pathetic face, jabbed a finger at the paltry, sad onions, and simply said, "Look." At least on the level of commodities, the imbalance in trade between China and India is clear. Sen's purported "unhindered" flow of goods and people along the Nathu-la route is anything but.

One of the many questions people ask about the rules for Nathu-la border trade is, Why the imposed limit to the exchange of goods at the border? Why only fifteen exports and twenty-nine imports? The answer may have to do in part with state-level competition, and the concern—at least on the Indian side—of surplus Chinese goods "flooding" the Indian market. As one trader in Kalimpong stated a couple of months prior to the reopening of Nathu-la in 2006:

> "Trade brings prosperity," they both [China and India] say. Of course there are other negative things, like security problems, but they [China and India] will put forces there. India is very cautious, because if they open their borders for everything immediately, the Chinese will dump their goods into India. Even small things like Chinese-made locks will kill the Indian industry, which has dealt in locks for ages. In Nepal, there is a quota, that way the local Indian interests are safeguarded, so it's not free-flowing. The same will happen through Nathu-la,

they will have a quota system, but the roads are not great right now. You have to build very good roads in order for this to work. This past season, all the roads were blocked from Siliguri and Darjeeling. It's still a very shaky thing. The infrastructure is not here, and lots of maintenance has to be done before more trade comes. Here everything has to go through Siliguri, so right now Nathu-la is for local trade only. The present roads are not capable of handling all the trade. People in Kalimpong feel sad that the trade was given to Gangtok, because in Kalimpong there is now nothing connected with the old trade. The trade is currently for Sikkimese subjects only, but India is guarding the interests of Sikkimese people. If it was opened elsewhere in West Bengal, big traders will come in. Quickly they will take advantage and Sikkim people would have lost out. But the Sikkimese will not be able to handle it when it gets bigger. It's an international border, so it will become international trade. It will then become Delhi trade. Maybe the big businesses will lobby to get a part of it, but that will come later. And of course the government will reply to the Sikkimese people who then complain that their local trade has been taken away, "Yeah, but look—we gave you a head start!"

It is not enough simply to point out that there is a contradiction between opening borders for trade while enforcing restrictions at these borders. Smaller-scale traders do manage to circumvent the restrictions in a number of creative ways, and despite the rigid government-led standardization of trading space and time, in actual practice, trade on the border is marked by the complexity of identities and places, creating confusion. In the next section, I draw attention to three thematic examples of place-making on the borders of the Lhasa–Kathmandu and Lhasa–Gangtok trade routes. As pointed out earlier, I choose not to question whether trade practices are always emblematic of "resistance" to power, for they often overlap with and contribute to the shaping of powerful state apparatuses. Although the border simultaneously unites and separates, traders' practices take the form of connecting, dividing, diverting, and reconnecting the route in specific ways that are often markedly different from the state rhetoric of what the route *should* look like for capital expansion. Trade routes are formed by various kinds of trading practices, both hegemonic and not.

Reconnections: Separating and Uniting

One border narrative that came up time and again was that of exiled Tibetans living in India and Nepal who, after hearing about the reopening of Nathu-la, planned to reconnect with their Tibetan families on the other side of the border,

FIGURE 20. Reopening ceremony, Tibet side, Nathu-la, July 2006. Pema Wangchuk.

family members they hadn't seen in over forty-four years. Likewise, Marwari and Newar traders older than sixty were interested in meeting up with Tibetans with whom they had had business and social connections prior to the 1962 border closing. A Marwari trader who traveled to the border in the late summer of 2006 told me, "The most amazing thing was the rejoining of families across the border after forty-four years. Some people were crying the whole day. One eighty-five-year-old man met with his own daughter of fifty-five years, I saw this myself." I also heard a story from a half-Tibetan, half-Nepali Kathmandu-based merchant, Norbu, whose Tibetan side of the family lives in Yatung (a Tibetan town 52 kilometers from the Indian border). In June of 2006, his uncle bought land in Yatung in anticipation of the economic benefits that the border reopening might bring. Indeed, while I was in Tibet, I heard many stories about wealthy Tibetans and representatives of government ministries purchasing land in Yatung as early as 2002 and 2003, in anticipation of the border reopening. Norbu says: "Five days ago, my uncle called my mother. He called her [and told her] to come to India, to Nathu-la. [He said] 'They are now going to open it, we can meet there' . . . We haven't met him for over twenty years. From Tibet to Nepal he wasn't able to get a visa, there were some problems. But on Nathu-la, in one day you can meet."

Similarly, a Sikkimese journalist who had covered the Nathu-la reopening ceremony for a local paper noted that "it was amazing to see families get together in Sherathang [the trade mart on the Indian side of the border] after forty-four years. They brought letters and gifts; they would get in touch so that they could meet on a certain day. There were lots of people who pretended they were traders so they could see their families." Although the question of *who* can cross over the Nathu-la border from the Indian side is, at the moment, limited to businesspeople with residency in the state of Sikkim who hold special trading permits, the categorization of "trader" can be malleable in the case of families wishing to visit their relatives on the Tibet side.

Since the Nathu-la border discourse is about trade (and potential future tourism) for the express purpose of boosting the "national economies" of China and India, these reconnection stories are not typically reported by official Chinese and Indian media outlets. Like the "reawakening" of the Border Baba described earlier, the reconnecting of families calls to mind the sensitive political history and cultural reality of the shared Tibetan/Sikkimese region straddling India and China. The movement of Tibetan refugees from China to Nepal and India in the late 1950s, the Sino-Indian border closure in 1962, and Sikkimese independence in 1947 until its merger with India in 1975 were all geopolitical events that contributed to spatial ruptures, disconnecting families

FIGURE 21. Tibetan traders, Nathu-la reopening, 2006. Pema Wangchuk.

that lived in the same region. Family reconnections make visible a different kind of trade route from the discourses that emphasize the benefits of new financial markets and national stability brought on by the reopening of Na-thu-la, conflating the economic goals of the reopening of the border with more affective goals of "Sino-Indian friendship" and blurring the rigid boundaries of trade eligibility established by the state: traders are for the most part defined as economic actors supporting the nation, not as human beings with exiled cross-border families.

In a small electronics shop in Sikkim, I met with three generations of Mar-wari Indian traders. I was first introduced to the grandfather, an elderly man in his eighties who recalled what it was like to trade over the passes in the 1950s. Also in the shop were his middle-aged son and a grandson who was busy ad-vising his many customers on which mobile phone network was the best value for efficiency. The grandfather and his son had both gone to trade at Nathu-la upon its reopening, and they showed me pictures of the trade mart, the border, and the plastic flowers being sold at Sherathang. They were very unhappy with the selection of goods and the prices on the border, and the Chinese goods were much more expensive than they had expected. They described their journey over the pass, noting that the grandfather, who remembered how to speak Ti-betan from his days of trade in the 1950s, wished to speak some Tibetan with the traders from across the border. As mentioned earlier, Tibetan was the main language of trade at the time. He noticed that all of them were speaking only Chinese, even if they were ethnically Tibetan. "Maybe they were worried," he wondered. He surmised that perhaps they had to speak Chinese because Chi-nese officials and guards were watching. He finally found a trader from the other side who was willing to speak Tibetan, but only privately in his van.

In this case, the use of the Tibetan language belies the official languages of the nation-states: in both the cases of the families reuniting and the use of Tibetan in the van, we see that ruptures or losses resulting from the physical dislocation of territory are met with the strong propensity to reconnect these spaces, to fix them anew. The seemingly mundane material practices produce alternative spaces where these reconnections might occur. In contrast to lit-erature on cross-border "flows," cross-border actors such as traders are not necessarily continually mobile; instead, they use mobile practices to maneuver around fixed state restrictions while at the same time deploying various fixing strategies to counter the state-centered discourse of free mobility and flows. We are thus reminded not to speak of mobility as a homogenous thing. Through this dynamic of establishing fixity against mobility at different scales, we can begin to see how state power is articulated and at the same time circumvented by informal actors. The next story also demonstrates how the practices of people produce routes and histories of trade; this time we shift from Nathu-la on the Sino-Indian border to Kodari and the Friendship Bridge on the Sino-Nepali border.

Khasa and Kodari: (Il)licit Rice Cookers and Rum

It is the middle of the monsoon season in Nepal, and I am trying—rather un-successfully—to wade quickly through the flooded streets in order to meet Jigme at his one-room flat. Jigme shares the flat with his wife, baby, and sister-in-law in Jorpati, a fifteen-minute walk from the main Tibetan quarter of Kath-mandu. He is a stickler for time, so when I arrive, I am glad I am not more than a few minutes late. Sturdy Chinese plastic sandals, worn by most people to better navigate the watery back alleys, are lined up outside the door. I leave mine with the lot, pull away the Tibetan door hanging, and greet those who are gathered on the mattresses on the linoleum floor. There are some relatives there whom I remember from before. Jigme is a very serious young Tibetan man in his early twenties, with a Sherpa wife and a newborn baby. Small, clear brown glasses are taken off the shelves, and his wife, Dawa, pours in the milky sweet tea. She seems to have a very bad case of boils, and the baby is coughing and looks unwell. I recognize the glasses; they are the same ones that are used in most *ja khang* (teahouses) in Tibet. While we have boiled potatoes with chili, I look up and notice bottles and bottles of 100-proof Chinese distilled sorghum or rice liquor (Cn. *baijiu*) lined up neatly along the shelves. Although I have known Jigme since I arrived some five months previously, he finally begins to tell me details of his own story as a trader.

He tells me he is a *sanno maanchhe* (in Nepali, a "small man," or a small trader) and began trading recently, in 2004, after having worked as a wool washer in a carpet factory. He was introduced to trading by his wife, whose family lives on the border of Tibet and Nepal. "At that time," he recalls, "we were not married, though she used to do the same circuit. She used to bring T-shirts and mosquito nets to Nepal. At that time we knew each other; she would bring the luggage from Tibet to Kathmandu, and I would help her sell to the shops." He would also help her go door to door in small shops on the Tibet side of the border, in the Khasa area, asking residents and shopkeepers what they might need from Nepal. Often they would take "hard drinks" from Nepal, especially *khukri* rum (Nepali rum) and cigarettes, both of which become illegal if more than a certain amount is brought across the border.

I point to the bottles of *baijiu* lining the shelves and ask Jigme how he brings this shipment across the border. There are two ways, he says. The first way is to simply carry them across. Both Jigme and his wife animatedly describe the various methods of transport and concealment that traders use to take the items across the border. The second way is to hide the goods in other items. "With *tsampa* [Tibetan barley flour], there are not any problems. They [the border patrol] see it and say, 'Oh, this is [just] *tsampa*,' then you can leave. 'Oh, this is just *tsampa*.' With *tsampa* there is no problem."

Tsampa, a Tibetan staple, is often mixed with butter tea and eaten in fist-sized lumps throughout the day to stave off hunger in high altitudes.[6] Mundane, everyday commodities like *tsampa* (or toilet paper, batteries, or mosquito nets, for that matter) almost always travel along the same routes as the (il)licit commodities.[7] As Carolyn Nordstrom has written, "popular media would have it that drugs travel a 'drug route,' arms an 'arms route' and computers a more cosmopolitan 'high-tech corridor.' In fact, shipping routes *are* markets . . . All manner of goods can pass among them," from the totally illegal to the completely everyday (Nordstrom 2007: 8). Illustrating this notion of routes-as-markets, transporting electronics across the border is also illegal, but Jigme notes that "sometimes we also brought rice cookers from there [Tibet]. But for rice cookers, they [the border guards] don't make a problem. I don't know why." This statement further illustrates the Marxian claim that social relations appear as if they exist between objects themselves, the objects themselves being differentiated by class. Mundane items such as rice cookers are associated with smaller-scale traders and waved on, as opposed to higher-tech electronics such as stereo systems or laptops. Similarly, at Nathu-la, a reporter who was allowed to travel with traders across the border stated: "Yes, there are restrictions on goods, but people from the Tibet side will bring thermoses and lunch containers full of tea and rice for lunch, then sell the containers on the Indian side. People also brought electric cookers and said they were 'gifts' for their family on the other side of the border, but then they actually sold them."

Another trader who travels between Nepal and Tibet says that it is easier for women to trade because they can say they are going "shopping." He says, "They buy ten rice cookers and find other women to go with them in the truck so it looks like they are each carrying one." Rice cookers are not considered suspicious by authorities; they are for domestic or household purposes and therefore considered "women's" commodities. These gendered differences in both the temporal and spatial patterns of border trade are further examples of how the abstract representation of a single trade route does not match up with the complex, lived reality of multiple routes created by the traders themselves.

Yulian Konstantinov has written about similar processes of (mostly women-led) trade between Bulgaria and Turkey, where items for sale are considered "personal belongings." This kind of trade, he says, is absent from reports on the economic situation of Bulgaria, although it is extremely common. According to Konstantinov, the characteristic feature of this kind of trade is its *strategic ambiguity*, where the traders can pass for tourists and the merchandise can pass for personal items (Konstantinov 1996: 762, my emphasis). In a study of Zimbabwean women traders, the majority of those who travel out of Zimbabwe to South Africa and Botswana to go "shopping" are working-class women. Such trading requires a sophisticated knowledge of bureaucratic procedures and finance and sometimes threatens male notions of proper female activities (Cheater 1998: 203, 208). Overall, though, upper-class women (e.g., airline employees) can bring higher-cost goods (e.g., televisions) across borders with little stigma, whereas working-class women dealing in such goods often get reputations as prostitutes, as threats to the state, or as dangerous women (Cheater 1998: 206). These increased divisions between "big" and "small" traders as well as increased gendering and classing of commodities are examples of the consequences of the changing—and uneven—trade patterns in the region.

Depending on certain trading regimes, the very same rice cooker can be seen as women's "shopping," turned into a "gift" for relatives across the border, or, in some cases, seen as contraband electronics. Commodities themselves can be marked by gendered assumptions that may be useful or detrimental for the continuation of trade. At the same time, border experiences accentuate gender and class divisions of trade, where Sherpa women who are considered "lower class" cross borders with relatively small quantities of contraband and have a very different experience with state authority than wealthier Newar women who travel in planes to buy higher-quality goods in Guangzhou.

Pradip and Identification

Amir owns a small shop on the busy streets of Kathmandu near one of the main shared-taxi stands. Amir always asked about my research if I stopped in, but there was never a lengthy conversation about it until one day, when I was on my way out of his shop to catch a shared minibus back to our flat. "By the way," he called out, "I have a good school friend, Pradip, who is a trader, a *thulo maanchhe* (Nep. "big man"), who is always traveling back and forth from

Kathmandu to Guangzhou. Do you want to meet him?" We quickly arranged to meet the next day at a casual rooftop restaurant, popular with some of the young elite businesspeople of Kathmandu. Pradip made a dramatic entrance at the door of the restaurant, pillion riding on the back of an assistant's expensive Royal Enfield motorbike, wearing an immaculate, tailored pinstripe suit with a purple silk shirt. He fit Amir's description of a *thulo maanchhe* to a T. We sat down, and he was eager to tell me about his business. Like most of the traders I'd talked to, he began with his family's trade history:

> **Pradip:** My father did this business [trade] too from Barabise [a Nepali border town]. Material, like textiles, with gold and silver—brocade—was brought from China. My father used to do this job for thirty-five to forty years.
>
> **T:** And what do *you* trade in?
>
> **P:** Anything. Everything. Whatever will sell. Whatever people need here.

Pradip's mobile phone rang. A shipment of paraffin wax from Saudi Arabia had come in, as well as a truckload of powdered milk from Khasa. While Pradip was on the phone, Amir remarked, "In older times, I still remember, they used to call Lhasa the source of gold [Nep., *Lhasa maa sun ko khaanii*]." Pradip rejoined us and said, "During my father's time, whatever they invested in, they can get three times profit. Anything. That was the 1950s or so. But now all things are coming from China. Everything is coming from China! Even *daal bhaat* [lentils and rice, a staple dish in Nepal and Bengal] is coming from China!"

Pradip showed me his ID card, which allowed him to travel to China. It read, "Foreigner Identification Card from Zhangmu/Dram/Khasa" (the Chinese, Tibetan, and Nepali names for the same Tibetan border town), valid only in China. "I have lots of passes," he said. He told me about some of the traders who live in the borderlands between Tibet and Nepal. "What they do is that they have a Chinese passport, but they have also taken Nepali citizenship in the area of Tatopani [a Nepali border town]." Pradip said that the traders are able to swap IDs when they reach the other side of the border. "Sixty percent of the people do that. They can travel anywhere in Nepal and also anywhere in China." Another trader from Lhasa mentioned the same thing about traders from Dram: "Their identities are like, they have two . . . most of them have two passports—a Nepali passport and Chinese passport. So they can just go back and forth freely and so that actually affords them a really good chance to be the middlemen, to take advantage of both sides. Buying stuff and selling stuff, without paying tax sometimes even. Because they know both the languages, these are the middlemen."

According to Sara Shneiderman, "The 1992 implementation of a Sino-Nepalese treaty, which allows citizens of either country who reside within 30 km of the border to cross freely without a passport or visa, has allowed many families to reunite. The provision has also proven an advantage to some families, who have been able to establish joint-venture businesses" (Shneiderman 2005: 32). However, traders must still carry a full passport and visa anywhere outside the thirty-kilometer radius, so having both passports, while illicit in certain areas, allows for a much greater range of movement.

In border zones, nation-bound identities are assigned, taken, and withheld; passports and papers become extensions of the traveler, defining who people are supposed to be, according to the strict criteria of the state. Traders' practices, including (literally) exchanging identities, demonstrate that border-crossing experiences are diverse and complex. And this practice of switching IDs extends to commodities as well; a businessman from Kathmandu mentioned that when he travels to Kodari (the checkpoint between Nepal and Tibet), traders who come in Chinese vehicles take the "Made in China" labels off mock brand-name clothes coming from Tibet and replace them with labels from other countries. The point here is that although labeling someone (or something) as "Nepali" or "Chinese" produces a standardized marker for the "authenticity" of goods or the identity of individuals, these "fixings" of place are considerably more malleable in practice. At the trade marts on both sides of the Chinese-Indian border at Nathu-la, for example, traders have said that "you can't tell who is Tibetan and who is Sikkimese, so we have nervous, guarded, apolitical conversations with each other, but sometimes realize we are actually from the same side!"

Traders Big and Small

Jigme has just returned from a small trip to the Nepal-Tibet border, and we are sitting on a balcony of a flat in Kathmandu, overlooking an alley that is quickly filling up with muddy water from the monsoon rains and floating, discarded boxes of Frooti mango drink. We are discussing the most common commodities he has transported across the border during his time as a small trader.

> **Jigme:** For the Nepali people, we bring toilet paper and canned juice from China, different kinds of chocolate, like that.
> **T:** You have to carry small amounts?
> **J:** Small, small amounts. So all the time we cross again and again and again and

again, sometimes [the guards say], "Oh, I saw you more than four, five, times, be careful!"

T: Can you make a good profit sometimes?

J: Yeah, if the army don't make a problem on the road. Of course we can make it, but a lot of times, they make a problem on the road and we have to pay tax. Sometimes they say, "Oh, I need this toilet paper," and they take the toilet paper, and then about 150 [Nepali rupees] is already lost, because they took it. We cannot tell them "why are you doing that?" We cannot tell them this, because we are doing our trade illegally.

T: Most of the shops around here in Boudha that sell toilet paper and canned juice, do they get it from the small traders?

J: Yeah, they get it from us. The big traders, they do around 80, 90 million per year, like that. But for example, we do about twenty to thirty thousand [approximately US$318] . . . around Boudha and that area, from traders like us, we're not like big men.

I do not wish to suggest that small traders like Jigme should be seen as constantly opposing larger capitalist networks and state restrictions, for small traders often do not wish to remain small. In an account of smuggling across southern African borders, for instance, Nordstrom quotes a local woman working for the UN, who says, "While the world talks about diamonds and oil, food and clothing are the real profits . . . The juncture where one can finally move from subsistence to capital ventures, is here—it is food and clothing that builds the houses, buys the cars, and launches the businesses" (Nordstrom 2007: 54). Jigme hopes that bigger opportunities will emerge if he keeps selling small things like toilet paper and chocolates. There is, however, differential access to networks of food and clothing between Lhasa, Kathmandu, and Kalimpong that has allowed certain small traders to become big, while others barely manage to pay rent.

There is a historico-geographical dimension to the movement of big and small traders across the trade route. Consider the changes in connections between the cities along the routes, outlined earlier in this book. In the heyday of Tibetan trade in the 1930s–1940s, the big traders were families from eastern regions of Tibet who had connections to the elite members of the Tibetan government and links with the monasteries that would finance the mule caravans that carried wool and musk from Lhasa through Yatung to Kalimpong. Wool was eventually sold to English and American buyers. Soon after the Chinese entered the TAR in 1950, the state took over the wool trade, and the big players became the Marwari and Newar traders who provided goods to the Chinese army in Tibet. These communities later had to reroute their trade to Kathmandu after the border between Tibet and India was closed in 1962. Since then, the majority of overland trade from China to Nepal or India went from Lhasa to Khasa to Kathmandu. However, the routes are changing once again due to the expansion of Chinese capital and changes in inter-Asian infrastructural networks.

To exemplify, I spoke with one young Nepali trader whose family has lived in the Mahaboudha section of Kathmandu for several generations. Mahaboudha is the section of town where one can buy cheap electronics, household goods, and clothes from China, where (according to some bigger traders) the "worst-quality" versions of many everyday commodities (such as polyester sweaters, kitchen utensils, and electric lights) are sold. Rahim, who had traded between Khasa and Kathmandu for many years, claimed he had seen things change rapidly over the past few years. He said:

> Khasa was a place where businessmen, big and small, would go to gamble, they could find anything, buy things, get information passed back and forth, girls, karaoke, etc. Businessmen from Mahaboudha would do everything by themselves, they would go to Khasa, do negotiations, customs, then bring stuff back. Now, in the past two or three years, it has become easier and cheaper for people not to buy things from Khasa, but to go to Guangzhou and other places directly in China [through Hong Kong or Beijing] to deal with middlemen, interpreters, then they buy goods from the warehouses or factories, then have them sent, shipped overland through Khasa to Nepal.

Another trader located in Siliguri, Sanjay, was adamant that there was nothing to be excited about in regard to the reopening of Nathu-la. Sanjay owned a shop chock full of rather elaborate-looking Chinese-made furniture. Walking into the shop, I was amazed by the collection of life-size naked female bronze figures, plastic dolls, pink flowery lamps, bunk beds, canopy beds, and a thing that was probably a phone. It was about the size of a US postal box, on top of a stand that looked a bit like an ornate bedside table. The item glowed with a lit-up moving waterfall landscape, with what appeared to be a receiver on one side and a dialing portion on the other. These were the kind of conspicuous items that take up more room than necessary and were probably reserved for certain upwardly mobile Indian families.

Sanjay had only been in the furniture business for two years. Four years ago, he was selling suits; he had to change merchandise because of too much competition. He said that there were about a hundred import houses in Siliguri that were importing items from China. "India has too much taxes, customs, excise, sales, et cetera—enough that you end up paying seven or eight times as much," he explained. "China is winning because they don't have such taxes." Every two and a half to three months, Sanjay flew to Guangzhou or Yiwu City, near Shanghai, to order more furniture pieces and brought back five or six containers full of furniture through the port of Kolkata. As to why he now imported from China, he showed me a coffee table and said, "Because people don't want to wait long for their furniture anymore. If you order the same thing in India, it may take months for them to carve, to nail together, to deliver, and all for the same price: 4000 rupees. People want easy, ready-made items now; they can just pick it up and go home."

When I asked him about the new border opening at Nathu-la, he said somewhat disinterestedly, "With Nathu-la, nothing will change. It's nothing important, they only sell things like wool. We don't do that at all, so it doesn't really concern us. We use the sea route—we can profit much more through this."

Trade is thus only easier and cheaper for *some* traders—particularly big traders who can afford an airplane flight to Hong Kong or Guangzhou and the cost of an interpreter-intermediary or broker (mostly English-speaking Chinese Christians who arrange overland shipping and take a commission), and who can tap into these Chinese networks. Unlike their forefathers and small-trader counterparts, who must travel with small loads of commodities from marketplace to marketplace, big traders who go directly to Guangzhou become divorced from their commodities, allowing a new kind of interpreter-intermediary to take care of the logistics of overland shipping. Whereas the small traders used to be the intermediaries in Khasa, this new kind of agent pops up in Guangzhou in the form of interpreters. Where capital goes, so do the intermediaries.

Like many small traders, Jigme soon found himself out of work. In 2006, the Nepali government introduced new border restrictions coinciding with the inauguration of the Qinghai railroad, the reopening of Nathu-la, and the tightening of Nepal-China trade relations. These new restrictions included a compulsory identity-card system that Jigme was not able to use on the border, as only Nepali citizens can apply for the card. As a Tibetan refugee, he does not fall into this category. Often after state security looks as if it might be threatened, new passport routines with newly standardized ways of defining identity and personal characteristics are imposed (Löfgren 1999: 7). Jigme became deeply in debt, trying to take odd jobs here and there, putting up flyers, helping out at a local school, or working for other traders who had obtained the appropriate ID cards.

Although some of these traders (including Jigme when he used to trade full-time) "clearly benefit personally, their actions often bring development to their communities. In areas where infrastructure is weak and governments can provide little in the way of social services, it is people like them who help rebuild regional industry, donate to health and education, and bring in critical resources for their citizenry. This is the ultimate contradiction defining the extra-legal: it can act as both harmful profiteering and positive development—sometimes simultaneously" (Nordstrom 2007: 99–100). Besides distinguishing between trading practices that result in positive or negative development in local communities, there remains the need to ethnographically examine how such practices—both harmful and not—might develop together. If the creation of trade routes is the result of the complicated and often contradictory processes of moving and fixing, or of opening and closing, it is crucial to pit these processes against the backdrop of more mainstream state-centered discourses on borders and trade. Following a larger-scale perspective, the next chapter examines recent major shifts in capital, transport networks, and trade between India, China, and Nepal.

CHAPTER FIVE

New Economic Geographies

> The age in which Asia was penetrated and developed from its fringes towards the center is drawing to an end. A new age is opening out in which the focus of development will lie near or at the center, and the effect of this development will radiate outward to the fringes.
>
> **Lattimore,** *Studies in Frontier Society and History*

Expanding Asian Infrastructure

Written more than a half century ago, Owen Lattimore's statement seems remarkably prescient. During my fieldwork in 2006, in addition to the reopening of Nathu-la, other major efforts to build up infrastructure and industrial centers in the "deep hinterlands" of inner Asia were well under way. As part of the long-term economic vision of the PRC to "Develop the West," plans were being made to extend the Qinghai–Tibet railroad to Khasa at the Nepal border, and perhaps even to Yatung at the Indian border, the last major town in Tibet before Nathu-la. There was also talk of extending the rail network to the disputed region of Nyingchi on the border of Arunachal Pradesh, where several road-building projects have already begun (Ramachandran 2008).

Furthermore, additional road projects aiming south through the Asian Development Bank–led North–South Economic Corridor will link Chinese highways with cities in Burma, Laos, Cambodia, and Vietnam, all the way to Bangkok. An oil and gas line is being built from Kolkata up to Yunnan through Burma (Fuller 2008). Another border pass, serving as a dry port, opened in 2006 over the Karakorum Highway between Xinjiang and Pakistan. At that time the road could handle only two trucks a day, but just two years later, the Sust Dry Port website boasts that four hundred trucks deliver goods from China each day. As of June 2011, a freight-train journey between China and Germany takes only thirteen days, compared with thirty-six days on the maritime circuit.

The push for infrastructural development in China reflects the need to open new markets for surplus commodities. As almost-classic examples of David Harvey's description of the "spatio-temporal fix," railways and dry ports are "opened" as geographical outlets during overaccumulation crises in order to provide spatial solutions, assuring that capital continues to flow across borders. Opening a border or introducing a railroad for the quicker transport of raw materials and tourist capital thus becomes a spatial "fix" both in the sense of a temporary "repair to a problem" and in the sense of a "fastening or a localization" (Harvey 2003: 115). As I have written in chapters 3 and 4, this kind of "opening" and integration—in the name of globalization and "free" trade—paradoxically reestablishes China and India as state powers by *reasserting* their national boundaries (Arora 2008). Consolidating and resecuring state power goes hand in hand with the capitalist processes that establish these fixes in order to survive, a notion I return to in the final chapter.

Rapid spatial expansion driven by capitalist accumulation results in the lived experiences of "compression," or the speeding up of time and collapse of space, and actually stands in contrast to other kinds of social processes taking place in areas that are less visible—at least from the perspective of nation-states. In media reports on the economic rise of China and India—and indeed in much scholarship on globalization and capitalism—it is perhaps too simplistic to focus on processes of expansion over contraction, mobility over fixity, speeding up over slowing down, enlarged trading networks over loss of connectivity with former trading networks, and monetary over barter economies, for these unequal processes go hand in hand. As a case in point, Tibetan herders were encouraged to celebrate the construction of the Qinghai–Tibet railroad, a railroad that actually made it faster to travel between the distant urban cities of Xining and Lhasa than from their own herding village to the nearest town.[1]

Focusing on the wider and faster processes conceals the trends that often occur in obscured places. These trends sometimes reveal just the opposite: fixity rather than mobility, or the use of barter in place of money. Spatial expansion is never felt uniformly, as it creates differentiation between those who have access to more hegemonic economic networks and those who find themselves struggling to compete against or become part of these linkages, or to survive in their shadow. Their narratives, mappings, and practices may, but generally do not, match more hegemonic stories of global expansion and development.

In this chapter, I attempt to contextualize traders' differential experiences of regional infrastructural developments with the larger picture of recent economic shifts across China, India, and Nepal, arguing that the traders' new economic geographies are both a response to and part of the new geographies of a heavily China-driven economy. As the Chinese government extends its capital development westward into Tibet and searches for new markets in South and Southeast Asia, former trading hubs like Kalimpong are bypassed. By looking at proposed Qinghai–Tibet railroad extensions and future border openings as opportunities for new market niches, many young entrepreneurs in the region are forging new trading networks, turning away from the Lhasa–Kalimpong and Lhasa–Kathmandu routes of their parents and grandparents and instead forging tighter connections with major industrial centers in southern China. By abandoning long-standing social connections, they are revealing a different economic geography than that of their parents' and grandparents' generation.

I explore accounts from several different groups of younger traders—Rahul and Rina, Gyan and Jeevan, Dawa, and even some traders who have now reverted to barter—showing how directional shifts in trade have simultaneously given rise to seemingly "successful" and "unsuccessful" experiences of exchange. While some traders' new paths have resulted in social connections in far-flung cities, faster communication of trade information, and diversification of trade commodities, other traders have found themselves needing to halt, slow down, scale back, or reverse their trading activities in order to survive, at the same time reasserting their local experiences of trade. I present these cases of unevenness as further examples of how spatial solutions to flows of capital can create new problems. This next section begins with stories from Rahul and Rina, two "successful" traders who have managed to take advantage of some of the numerous new outlets for commerce through Asia.

Rahul

Toward the end of my fieldwork, I ran out of business cards. Talking to merchants and traders necessarily brought with it the formal exchange of such cards, and I had underestimated the number I needed. In Kathmandu, a friend recommended a stationery shop where I could get some printed up within a couple of days on handmade Nepali paper. The proprietor, a large, pleasant man whose brother was working on a master's degree in sociology, noticed my university affiliation and wanted to talk about my research. Upon learning that I was conducting research on trade, he said he would introduce me to a friend of his, Rahul, who was a *thulo maanchhe*. It was toward the end of the day, so he made a quick call to his friend, closed his shop early, and very kindly motorbiked us out beyond the Ring Road, the main circuit around the city. On our way out there, the stationer explained that Rahul's father had been involved in the old Tibet trade, and Rahul, in his early thirties, was now making it on his own as a businessman. We arrived in a section of Kathmandu I had never seen before; it immediately reminded me of the warehouses lined along the desolate navy yards of Brooklyn or Long Island City in New York.

The complex of three- and four-story warehouses was clearly a man's world. I kept my head down, wondering what people would make of seeing an unfamiliar, slightly foreign-looking woman being led up to the boss's office. Following the stationer up a series of many narrow concrete stairs, I looked out the windows to see porters transporting sacks of what looked like flour on their heads to the trucks waiting below in the late afternoon heat. Rahul sat at a desk in a far corner of the spacious room; it was sparse, with a landline phone and little else, save for a filing cabinet. He was talking on the landline with one hand while holding a mobile phone in his other, and I noticed at least five people, clutching wads of cash, standing in line to talk to him. I thought it would be a while before I could be introduced, but he beckoned me in, and I sat on a chair while the first man in line presented him with a stack of hundred-dollar bills.

Rahul told me that he had begun his own company fourteen years earlier by using his father's old business contacts, exporting Nepali wheat flour to India. As he began to phase out his father's flour business, he was quickly becoming more prosperous. Rahul had started importing raisins from Xinjiang, dates from India, and a variety of foodstuffs from other locations around the world. He said he was looking forward to taking advantage of the new railroad to Tibet, as he claimed it would result in cheaper prices of goods coming through China:

> **Rahul:** Now like fifteen days back, they have started that railway. That's it. We . . . hope that the price will be cheaper due to the railway. They send [goods] to Lhasa, and we will import them from Khasa to Kathmandu.
>
> **T:** How is that going to affect your business, do you think?
>
> **R:** It will be much better, much cheaper. I think . . . we are doing an export business, but we have to change to do imports. That will be better. Because, due to that railway, all the products will be cheap in China. That is why we cannot export anymore. Better that we import. We used to do export business, but now I think import business will be important. Importing from China and exporting to India.

Rahul's emphasis on having to "change to do imports" was typical of many of the younger-generation traders in both India and Nepal. Phasing out exports and relying on imports from China reflects a major shift in the geographical direction of trade in the region; as one elderly trader who traded in the 1950s put it, "In those days, we took the finished goods [from Nepal and India] and brought them back [to Tibet] . . . Now it's the other way around. From China you get electronics and TVs." As production centers shifted to southern China and easier access to transport networks between China and other parts of Asia drove down the prices of Chinese goods, one way traders in Lhasa, Kalimpong, and Kathmandu responded was by changing or diversifying the commodities they dealt with. This change in turn marks a shift in the traders' relative location. In Rahul's case, for instance, selling flour from Nepal was simply not profitable anymore, as Chinese flour could be obtained much more quickly and cheaply. Since he knew he lacked the assets to sell Chinese flour competitively (partly because he did not possess the linguistic ability or social access to Chinese flour supply networks), he cut off his connection to Nepal and diverted his trajectory of goods, beginning to sell products that were produced in locations outside of China. Rahul continued by saying:

R: I have started [to import] from Pakistan, Afghanistan, Sri Lanka, Singapore, Malaysia, Thailand, Vietnam.

T: Wow.

R: Recently, I have imported black pepper from Vietnam. Vietnam is one of the cheapest . . . black pepper from Vietnam we can get cheapest.

T: What kinds of things from Sri Lanka?

R: Do you know coconut?

T: Coconut, of course.

R: I've imported coconut from Sri Lanka . . . I am trying to . . . do you know gram [legumes]? Like peas. I have also imported peas from Canada. And I am importing from California also. Almonds . . . Nowadays I am starting [to trade] from all ways. I am importing California almonds. Almonds you know.

T: Yeah, I love almonds.

R: That one I am importing from California. And nowadays, after fifteen days, fifteen or twenty days, they will start a new crop. We have to get all this information from Singapore.

T: Information from Singapore?

R: From Singapore, we get information like, "A new crop is starting this season."

As Wolfgang Schivelbusch demonstrates in his intriguing sociohistorical account of the emergence of rail travel in the nineteenth century, if infrastructure's product is a change in location, then new localities and destinations often end up becoming commodities, too (Schivelbusch 1986: 197). Not only do some localities have more value attributed to them than others in certain circumstances, but some traders also have the resources to switch locations to take advantage of these new opportunities, while others do not. In this particular case,

Rahul increased his access to global information networks and a diverse range of commodities in order to remain competitive in a market now dominated by China, forging economic links to new places like Canada or Sri Lanka—new nodes of trade that did not even figure in his father's economic geography.

After I asked Rahul to tell me a little about where his father used to travel for trade, he replied, "He used to import [to Nepal] from India. [He traded between] India, China, and Nepal . . . and some of the items he exported out to Tibet, *but not internationally*" [emphasis added]. For Rahul, "international" trade involves dealing with people in places such as Singapore and Canada, outside his father's long-standing route, which, according to him, is merely "local trade." Of course, from a state-centered perspective, this older, "local" route is indeed international, as it crosses national boundaries and is subject to international customs laws. Rahul's claim that his father's trading geography was "not international" is firmly rooted in a kind of geographical representation that belies the state-centered logic of "national economies." Part of expanding his business to new locales involved making sure consumers knew his products were attached to specific places that now had considerable value because of the geographical distance of their provenance, such as almonds from California and pepper from Vietnam. Although increasing mobility and travel is often considered one of the hallmarks of globalization, the story that is occurring here is one that involved Rahul staying put while the goods continued to move. Rahul found that he made only occasional, quick trips abroad (in contrast to his highly mobile father's generation); he now mostly remained at his desk, making phone calls to Singapore.

Rina

Another, similar story concerning the generational expansion of trade involved three Kathmandu-based traders from two generations—Daya, an elderly man in his eighties, and his daughter and nephew in their forties, all members of a Newar family currently involved in the hotel and retail business. They spoke frankly about the spatial differences between three generations of traders. According to Daya, his father didn't travel along the Lhasa–Kathmandu–Kalimpong trade route; he would simply go to rural Tibet and buy woven woolen items (chubas and aprons) direct from the villagers and bring them back to sell in Lhasa. That was the extent of his economic geography. Daya, on the other hand, was part of a family of five brothers, all of whom took turns conducting business in Lhasa. In the 1940s and 1950s, they each spent four years in Lhasa, alternating with two years in Kathmandu or Kalimpong, importing items that arrived via the ports of Kolkata or through Kashmir, such as coral, pearls, turquoise, and fabrics, and exporting musk, yak tails, wool, raw gold, and silver from Tibet. Now, over the past few years, Daya's daughter Rina had begun to fly to Thailand and Guangzhou at least four or five times a year to pick up new items for the supermarket the family established in the late 1980s; she also ordered gourmet items for wealthy foreign expats and Nepalis, such as French brie, American facial cleansers, and Australian milk in cartons. Rina noted that during her father's time, trade was "strictly a family business" associated with the Newar ethnic group, "but now, you find all sorts of people coming to Guangzhou from Nepal; there are all sorts of people doing trade."

Both Rina and Rahul's stories reflect the classic, hegemonic "success" stories of globalized Asian trade that I heard throughout my fieldwork, where successful traders are considered those who forge new business contacts with individuals living in far-flung locations, and whose imported products come from increasingly distant places. What often gets neglected in the celebratory discourses of the "new" Asian trade, however, and what I wish to show here, is how the inherent unevenness and contradictions of geographical development are both experienced and produced, and how multiple countermovements of place-making are in tension with state-based moves and success stories. Turning to a less-fortunate group of traders, we see how competition due to changes in the geographies of infrastructural networks slows some down.

Directional Shifts: New Goods, New Geographies

Having a stack of business cards, as mentioned earlier, is integral for traders needing to make new contacts with suppliers, buyers, and distributors. They can also be indicative of the changes in the geographical direction and social restructuring of businesses. In Kathmandu, I was handed a business card with "Om Shanti Wholesalers: Retailer and Wholeseller Order Supplies for All Kinds of Chinese Goods" written in English on one side; on the other, "Worldwide Marriage Bureau Nepal: Your Life Time Partner." The proprietor grinned and told me about his venture to create an online international matrimonial service: "We needed to expand," he explained. In Lhasa, I was handed a fresh-from-the-press business card for a new Taiwanese jewelry shop with Chinese on one side, Tibetan and English on the other. In Kalimpong, I was given a business card that said, "grocery, provision goods, and miscellaneous items." But that phrase was crossed out in ballpoint pen and "MOBILES" was handwritten in, in capital letters. In Siliguri, I received a card that read, "Electronics: Everything in Electronics; Automobiles; Imported Furniture." And in Sikkim, one man (who used to deal in wool) gave me a business card that advertised his supplies of Sony TVs, washing machines, mobile phones, DVD players, and toaster ovens. Clearly, traders who have been able to tap into the new economic networks linking North Bengal, Nepal, and Tibet with inland China now profit from selling Chinese-made products or from providing services that are not physically tied to the region, such as online international dating websites. Like the changing commodity items and fluctuating prices listed in nearly every issue of the *Tibet Mirror* newspaper in the mid-twentieth century, business cards can act as material indexes of the paths of people and goods, as suggested by their diverse mappings: Siliguri on one side, Yiwu City in China on the other.

Gyan and Jeevan

Gyan and Jeevan are two cousins in their forties and fifties who said they had to create different sets of business cards each time they began to sell new products. They were some of the very first Nepalis who returned to Tibet in the 1980s to continue their family's trading business after the Sikkim–Tibet mountain passes closed in 1962, reorienting the majority of small-scale Tibetan trade to and through Nepal. The PRC's economic reforms in the mid-1980s—which included attracting foreign investors to port cities, establishing Special Economic Zones in locales like Shenzhen, and opening up the TAR to tourists—were what chiefly prompted Gyan and Jeevan's move back to Lhasa, their fathers' town of business. They recalled what the economic atmosphere in Tibet was like at that time:

> **Gyan:** In 1985, it was very easy to do business. No road came from China to Lhasa.[2] We could supply everything [to Tibetans from Nepal]. You know, rice, sugar, wheat flour. Everything came from here. Medicine, herbs. Everything! You could supply to Tibet all the things . . . Bricks also, cement. And sand also! You could supply. They can build the buildings. And vegetables. You know, at that time, it was very good business. We could send the vegetables in one truck, and they can barter. They gave us *byi ru* [coral] and *g.yu* [turquoise]. They can give. I was very happy. One truck, they gave five kilograms, ten kilograms [of stones]. At that time it had no value. But *gzi*, you know *gzi* [a precious Tibetan onyx stone]? That time, they can give more and more. We didn't know [pretending to pick up a bead and examining it]. What is this? (laughs).

T: What did you do with it?

G: We can use it as money in Kathmandu market. At that time, we didn't know [what it was].

Gzi beads are highly coveted for their protective properties in inland China, and many traders now believe that there are no more real *gzi* or coral to be found in Tibetan markets. Gyan and Jeevan said that as traders, they were rather ignorant in the 1980s; they sold a lot of goods, but could have made a lot more money later on if they had known exactly what they were dealing with. In the late 1980s, when other Nepali traders began to get involved in the Tibet trade again, the cousins realized that they were beginning to lose money due to increased competition in Lhasa. Taking advantage of Tibet's opening up to foreign tourists, they began bringing Nepali handicrafts (such as homemade paper products with Tibetan symbols printed on them) to sell in Tibet.[3] At the beginning, their business fared well because the routes for transporting goods from Nepal to Tibet were easy compared to other routes in China or India. In the mid-1990s, though, Gyan and Jeevan had to halt their handicraft business in Lhasa, citing two major reasons: competition and corruption. More than fifteen years later, they noted that the exact same handicrafts they had sent to Lhasa from Nepal remain on the shelves. As Gyan said: "Now there's very much competition, now I cannot sell Nepali handicrafts, [because] everybody brings them illegally. We are bringing legally. But our price is very high. In 1990 I sent goods there [to Lhasa]. They are still there! I cannot sell. They are still there. It's very difficult to do business now . . . That's why [their company name] now stopped the business. In 1985, everybody knew [our company] . . . Now people, from different places they come. We don't know who is who."

The decline of their handicraft business occurred around the time the influx of Chinese migrant workers into Tibet reached its peak in the mid-1990s; Gyan and Jeevan therefore decided to switch to yet another product they thought would be viable. They became the suppliers of steel rods and cement for construction projects in rapidly developing Lhasa. As steel rods are cheaper to manufacture in Nepal and to transport from Nepal to Tibet than from other parts of China, they estimated that these items would be sure bets for a quick profit. In 2006, the cousins visited Tibet for a China-Nepal trade fair but returned to Kathmandu, disappointed. At the fair, they were hoping for "actual business" but were concerned that the Chinese traders "were just talking for future business." According to Gyan, "They said they will take much quantity . . . one year, one billion tons. And we bring a sample there, and [it was] only talk! Not actual business . . . it's difficult to deal with Chinese people now. It is very difficult." Gyan and Jeevan also lamented that the Chinese were now using inexpensive domestic cement for construction, not the Nepali cement that the cousins used to supply.

With materials like cement becoming cheaper to produce, purchase, and transport from within the PRC, not to mention improvements in domestic infrastructure in China, traders such as Gyan and Jeevan struggle to keep up with rapidly changing geographical knowledge by expanding their networks to new locales or switching to more competitive products from different places. Although their networks might be expanding geographically—from a village in Nepal to a supplier in Singapore—a certain element of face-to-face business is felt to be lost. As both Rina and Gyan expressed, they "don't know who is who" anymore. Similar to the narratives of decline discussed in chapter 2, capitalism presents a historical pattern of an increasing sense of alienation and the decline of trust as new, fleeting relationships displace long-term trade partnerships. Part of this sense of not knowing "who is who" anymore also results in a turn to a cultural argument for economic problems, sometimes slipping dangerously into anti-Chinese sentiments, exemplified by comments like "Chinese people are difficult to deal with," or, at other times, into ethnic self-deprecation or self-criticism, as explained by another Nepali trader living in Nepal:[4]

> No pain, no gain is what I always say and tell my wife. We have to learn to make money too, it can't only be the Chinese, but now it is. Only the Chinese are making money in Tibet. What about us? The most dangerous thing is now Chinese come to Kathmandu to settle and do business. Let me give you just one example: apples. Nepalis used to sell them, but when the Chinese started taking over the market and moved here, they had very good connections with suppliers in China and they make one phone call and can get the whole shipment here quickly. Nepalis don't know how to compete. They can't compete . . . we used to have the monopoly in Tibet, but not anymore.

This sense of alienation and lack of trust was noted by many of the mid- to large-sized traders in Lhasa, Kalimpong, and Kathmandu who were feeling hampered by increased competition and the dominance of traders directly linked to the paths of new Chinese capital (recall also the story of Tsering, the *pang gdan* workshop owner in Lhasa portrayed in chapter 2). "Opening up" spatial outlets to increase the flows of commodities sent from production centers in China yields a rather different experience on the part of some less-successful traders in Kathmandu, who feel cut off from interactions with other traders. In order to further draw out the experiences of traders who feel the spatial dimensions of a lack of sustained face-to-face interaction, I next turn to a story that demonstrates another kind of "geographical expression of the contradictions of capital" (Smith 1990: 152).

Geographies of Barter

There were very few tourists in central Kathmandu in April 2006. The beginning of the trekking season is normally a popular time to visit Nepal, as the skies are relatively blue and clear prior to the heavy summer monsoon rains. During this April, however, the city was in turmoil. King Gyanendra—whose takeover of the government, complete dismissal of the Nepali parliament in 2005, and subsequent military rule sparked major protests among most citizens—had declared day-long curfews to curb the opposition. The US State Department had issued an order against unnecessary travel to the region, and other foreign tourists were canceling their planned flights to Nepal. Tensions were very high, especially since staples like gasoline and cooking oil were in great demand.

I sat in a small jewelry shop with a Tibetan Muslim family I had met when first visiting Nepal as an undergraduate in 1997. Salima, the owner, and her daughters Fatima and Aisha chatted while Fatima strung together small pieces of turquoise and fake coral to make earrings. Usually by that time of year, the streets would have filled up with tourists and hawkers. No one showed up all day, and to say business was bad that year would be an understatement. We ordered meat dumplings and intestines for lunch and ate them atop the glass cases of stones and jewelry. Peering in the case and admiring the piles of rich blue lapis stones, I told Salima that I knew that lapis came from Afghanistan or Iran, but how did she actually acquire them? Salima responded that men from Afghanistan came to Kathmandu with bags of lapis lazuli, and in exchange, she gave them silver necklaces or rings. Because they take these things back home to Afghanistan, "of course we don't really give them the Buddhist stuff like reliquaries. Sometimes we exchange, and sometimes we give them money," she said. I was intrigued, since almost all of the traders so far had told me that they did not barter anymore, that barter was something that their grandparents did in the "old society." "Sometimes we don't have [enough] money," Salima said, "so you [ex]change." Her daughter Aisha added, "It's not like we have cash available anytime, so here you don't need to give cash." After this incident, I began asking more traders about barter. In Kathmandu, a man who sold *malas* (Buddhist prayer beads) said the following:

Yeah, yeah, because people have not much money, they do like this. Tibetan women who come from Lhasa, when they [can't] sell their things, they exchange with other things . . . If they cannot sell, if they need something, then they exchange. By force they do barter system. Many women come here to the shop, they exchange with *mala*. If they need something like *mala*, and if they didn't sell their things, they need these things, and if they also have no money, then they [exchange]. That's by force! . . . By force! They are doing this still. Before there was no money system, and people did the barter system. Even today in the modern age, people do the barter system—but by force. If I bring this [he animatedly pulls down objects such as bowls and plastic necklaces made with fake coral] from Lhasa, [sometimes] I cannot sell it in Kathmandu, because there is no business. I sell very few things, but this is a big town. I sell some, but I cannot sell [everything]. Then I take this in return. Then I go to my friend. "You exchange with this." I exchange with this [showing me some earrings made of silver and turquoise] in Patan. And I exchange that thing with one football with one boy, and he gives me like a Chinese pot, it's made with marble, it looks nice, they make it cheaper. Then I exchange with him . . .

I take the use of the phrase "by force" to mean that they are forced to barter, that they have no choice in these circumstances. People barter when it turns out to be a more viable economic alternative. In Kalimpong, for instance, I went with my Tibetan friend Lhamo after lunch to go and meet an eighty-four-year-old Marwari trader, Ramdev, and his grandson. In the old days, Ramdev said, wool from Tibet would be bartered with ready-made clothes sold in Kalimpong. When I asked, "But people don't barter anymore now, do they?" his grandson replied, "Oh, of course they do, they do all the time!" Lhamo, nodding, agreed. These days, they exchange patterned silk from China with sacks of Kalimpong-made *kha btags* (Tibetan ceremonial scarves).[5] Norbu in Kathmandu says that these Kalimpong scarves are of lesser quality:

The nice *kha btags* [100–200 Nepali rupees (US$1.40–2.90)] come from Lhasa [a big markup from 4–5 yuan (0.50–0.63 cents)], and the lesser quality come from Kalimpong. Merchants come from Kalimpong to the shop to sell, and back in Kalimpong, they still barter these things for what they need . . . The [Lhasa] *kha btags* are taken to India, [through] the barter system. If the Indian or the Nepali people, they bring Chinese goods like *kha btags* or something like that, they mostly exchange with goods like an offering bowl, or something Indian like a necklace, they'll bring here . . . if most of the Tibetans are poor, they come here,

> if they do not have money, they can exchange. Some have turquoise, some have *mala*, they can change with mostly *bum pa* [vases], butter lamps, offering bowls.

In one early, influential study on cross-border trade in the western Himalayas, Christoph von Fürer-Haimendorf predicted that because "Indian commercial influence" has reduced the scope of the Sherpa salt-for-grain barter system and because the area was "more and more drawn into the monetary economy," bartering would "soon vanish from memory" (von Fürer-Haimendorf 1975: 5; 3). Fürer-Haimendorf's assumption that there is an essential division between monetary and nonmonetary societies, or that a monetary economy "arrives" and erases a nonmonetary economy, is not quite accurate, as evidenced by the stories above. More significantly, perhaps, wider hegemonic economic geographies (such as "Indian commercial influence" in Fürer-Haimendorf's study or Chinese development initiatives in more contemporary accounts) might not limit barter practices but in fact ensure their continuation. So what might the use of barter mean now, spatially?

Barter has often been seen by traders and scholars either as a "lower level" of exchange in a hierarchy of economic practices, or as associated with a romanticized "egalitarianism" of the past. Instead, barter can occur at the same time as monetized forms of exchange, making special sense in contexts when cash is unavailable. In fact, "barter should be seen as one mode of exchange amongst others, not as the single means of running an economy" (Humphrey and Hugh-Jones 1992: 6), a point that David Graeber also makes when challenging the popular notion that barter somehow begat monetary exchange and credit in a linear progression (Graeber 2011). Meanwhile, monetized exchange itself has become more diverse. As "successful" or "big" traders like Rina and Rahul increasingly use faster modes of financial transactions, phoning up their contacts in Singapore and Bangkok to order almonds from California and brie from France on credit, their economic networks become more far-flung. Unlike barter, these faster, credit-induced processes cause the transfer or exchange of money and products to become much less visible.

There is credit and there is credit, however, so much so that stories about the use of credit also reflect differential relationships to trust and the temporality of exchange in the new economy. Interestingly, relatively well-to-do businessmen such as Dawa (recall his story, in chapter 2, about trying to make Tibetan handicrafts a "trend") said that purchasing items on credit was done only in the past:

> **D:** I think it was common before to take a long time, to take a rain check, to wait like months. And it was even easy to get loans, like personally, between

friends, or even from the neighborhood. But now, it's tough to get loans, and there's huge interest involved. People are just crazy about securing their selves.

T: Are there forms of credit now?

D: Um, very few. I mean, still credit is involved, but it's a pretty unusual thing right now. It's like cash, cash all the time. Immediately. Seriously. And it's got to be real cash. Not duplicated, fake. It's crazy today. I guess that's . . . people are just forced to be like that right now, because of the environment. The social environment is just so . . . the whole environment is just crazy.

Some small traders who are struggling to make money say that they will delay exchange and sell items on credit—but only if they trust their customers. Take for instance a thirty-year-old Sherpa woman in Lhasa who managed a small cosmetics stall and whose husband traveled to Nepal every month to purchase new supplies. When she sold Indian and Nepali skin lighteners and shampoo to women from other urban places in Tibet such as Chamdo, Shigatse, and Gyantse, she said that since they didn't know each other, the customers must pay right away. In her corner of the main marketplace in Lhasa, however, since most people knew and trusted her, if they didn't have money, she would let them pay her back later, between ten days and a year. When a friend of hers bought some Nepali butter, he didn't pay her for five years, while he was in the United States.

A former wool trader also mentioned that it used to be more common to take a long time to decide what you wanted to purchase, because "people were sort of more trustworthy before. More honorable. And less decisive, I guess." He said that there was something particular about barter that enabled the transaction to take a fairly long time, that "because there is no standardized sort of price in many cases—basically, the trading has to be negotiation. It takes a lot of negotiation, as far as the price goes. You have to decline the price as the buyer and for the seller you have to keep it profitable. You don't even know what the profit margin is sometimes. They will never calculate that, but you know, that's the way it is. Have you ever seen people bargaining without using their mouths? Just by their hands, it's kind of confidential—that sort of reflects the secret business." I said that I had, especially in the Barkor, near the vegetable market, and had met some of these traders. This kind of silent trading—called *phu dung nang tshong*, or trading inside sleeves—is a Tibetan trading practice conducted entirely under long shirt sleeves, where those who wish to conduct business take hold of parts of the hand in order to communicate certain prices to each other. Two sisters of some Khampa traders once tried to demonstrate these hand signals; one women grabbed my thumb and said it was the *a ma* or *pa lags* (the mother or father) and stood for 10,000 RMB; then she grabbed my pinkie and ring finger together and said they were called the *bu* or *phru gu* (boys or children) and together represented 2,000 RMB. Altogether, the transaction she was asking for was 12,000 RMB. Several semiprecious stone sellers from Kham said that this sleeve-trading practice was still very strong. It also remains prevalent in the contemporary trade of caterpillar fungus (*yartsa gunbu*, a high-value product used in Chinese and Tibetan medicine), although some people in Lhasa say that it is dying out or that it was done only in the past. Such trading practices definitely occur alongside and in addition to cash-based transactions, especially when the goods are of high value, like old coral or caterpillar fungus, and are familiar even in situations involving high-level finance capital such as the trading of cattle futures in the Chicago mercantile exchange (where the hand signals maintain the confidentiality of clients but are made visible). Sleeve trading is an example of an older practice that still occurs between certain traders, while others—particularly those on their way to becoming bigger traders, at least in the Chinese sphere—are learning how to participate in other, less-familiar Chinese trading practices or etiquette. The next story is about this transition.

Stetsons and Cigarettes

Dawa, the young man who was interested in setting up a handicrafts center in order to revive wool for a new generation of upwardly mobile Tibetan consumers, had been explaining how he had to "learn what it meant to do business" in rapidly developing Lhasa. As a Tibetan who had lived in Nepal and dealt previously with Tibetan exile, Nepali, and European business contacts, Dawa poignantly described how he had to learn to approach local officials—both Chinese and Tibetan—when planning to purchase land outside Lhasa in order to build his new center. He was told by a Tibetan friend who spent much of his time in mainland China to take the officials out to dinner to begin the initial negotiations. Fair enough, he thought, and he told everyone to meet him at 7:00 p.m. at an upscale Chinese hotpot restaurant on what is colloquially known as "Thieves' Island," a section of Lhasa famous for its nightclubs and restaurants, and where business deals often take place. What Dawa didn't know, however, was how to do business, as he put it, "properly, according to the Chinese custom." This involved booking a special private room at the restaurant, heading there at least half an hour ahead of the scheduled meeting time, say

6:00 or 6:30, and making sure there were at least three sixty-five-yuan boxes of cigarettes for all of the attendees at each place setting. As Dawa sauntered up to the restaurant door at a quarter to seven, his friend was already waiting, pacing back and forth. "Have you booked a room?" he asked. Dawa had, but the room wasn't deemed adequate enough by his friend, so they quickly negotiated with the staff and moved to a more elaborate-looking room. "Did you get the cigarettes?" his friend asked. "What kind of cigarettes?" Dawa's friend, now in a panic, quickly rushed out to get the required top-shelf brand cigarettes, then took complete control of the "proper customs," and the rest of the evening ran smoothly. Later on, the officials were taken to a karaoke bar where there was much drinking — so much so that Dawa had a difficult time handling it. Dawa's wife, much more familiar with the Chinese business world due to her time as an office worker in Beijing, laughed knowingly when he returned home, exhausted and tipsy. At one point, Dawa told her he was simply going to forget buying the land because of all this "hitting the bush" (this interview was in English; I assumed he meant "beating around the bush"), but his wife insisted, "No, you must press on, press on." She even took over for him at a negotiation dinner one night when he was ill.

Dawa rolled his eyes, explaining that there was also the custom of what he called "having to buy things for people in order to get things." Some examples of purchases were the latest digital cameras, "or cell phones you can't get here in TAR." This involves what Dawa calls "giving a little taste." "It is like giving a gift but not," he explains. "It involves saying things like, 'Oh, have you seen this new digital camera? It's pretty nice.'" Dawa mimics holding out a camera to me. He says that if the people you are doing business with seem to admire it, then you say, "Oh, well you can keep it," and then give it to them. US brand cowboy hats, according to Dawa, are another desirable commodity that one cannot get in Tibet. This is true. Cowboy hats in general are popular among both men and women in Tibet, providing good covering on the sunny and dusty Tibetan plateau. High-level Tibetan officials or well-known performers who appear on television during special programs where they must wear Tibetan clothing don extremely nice cowboy hats, sometimes Stetsons that sell in the United States for eighty to two hundred dollars. Once, when he was in New York City, Dawa went to JJ's Hat Shop on Fifth Avenue, a famous old NYC hat establishment, to buy fifteen cowboy hats for officials, and the elderly black proprietor in a fedora and three-piece suit simply couldn't believe that Tibetans wanted cowboy hats. Dawa told me that he explained to the man at JJ's that "the US has the Wild West, and Tibet is kind of like the 'Wild East.' Even though they don't wear Wranglers, and they ride smaller horses, it's pretty much the same."

He also purchased a few crushable cowboy hats that pop back up again, which were particularly popular with bureaucrats. All in all, he became much better at learning how to do business, and the officials were very impressed with him.

Dawa's "business education" in the rapidly developing and competitive business atmosphere in Tibet—the frontier land of China—was strengthened and cultivated by his wife and several friends who either had grown up in mainland China or were already familiar with doing business there. At first he had no idea how to go about giving presents "so that the other person doesn't get embarrassed that you are giving a gift, but understands what you are doing and accepts it." Shaking his head, Dawa says that such things are much easier in Nepal, where "at least they are straight with you when they want something in return. They say, 'Give me baksheesh,' and then you do, and the deed is done. Here it is just so difficult to understand how business works. I am always 'hitting the bush.'"

New paths of capital and investment in Asia are producing extremely different kinds of spatially relevant exchange patterns. In Tibet, barter and credit transactions conducted "by force" as a result of unevenness in capital flows remain a way to survive for some traders. Such transactions rely heavily on creating visible personal networks and selling specific items that are immediately available for exchange. Although traders like Rina and Gyan complain that "no one knows who is who" anymore, some of the less-successful traders who resort to barter do this "mostly in face-to-face situations, where people and the paths of goods are known" (Humphrey and Hugh-Jones 1992: 6). Turquoise might be exchanged for an offering bowl, for instance, when money is unavailable; such a practice happens in a very specific time and place, nor does it always trump the use of money on other occasions. Thus, when money is limited, these traders tend to foster more spatially compressed trading places, creating a different kind of economic geography that often gets neglected as a kind of geographical obfuscation of its own. Trading practices such as barter are not often "seen" by scholars who are intent on characterizing the recent increase in global connections and communications in Asia as a phenomenon driven purely by monetary processes. Although it might seem as if time and space are compressed for Rina and Rahul, who have more immediate access to expanding global economic networks, in the case of the examples of barter, opposite processes appear to be occurring. Although Salima knows the Afghan lapis trader by name, it takes much longer for the Afghan trader to return on his own to Nepal with new goods than it does for Rahul to receive a new shipment of nuts from Singapore, causing the Afghan trader's particular circuit of commodities to be temporally expanded relative to Rahul's circuit. Cindi Katz has recognized these processes of time-space expansion in her work in Howa, rural Sudan. After the area south of Howa had been deforested due to the northern war effort, people in Howa found that agriculture and animal husbandry were viable only if they traveled longer distances to graze and to get wood (Katz 2001: 1224). As Katz argues, "From the vantage point of capital, the world may be shrinking, but, on the marooned grounds of places such as Howa, it appeared to be getting bigger every day" (Katz 2001: 1225).

But what do we make of Dawa's story? He is conducting business in this transforming Asian economic landscape; in order to purchase land in a developing Tibet, he must learn rules that are new to him, some of which involve understanding how to have good *guanxi*, or personal connections, and some of which involve gift giving, favors, or even exchange processes bordering on bribery. These practices run side by side with other forms of exchange and are tied to infrastructural shifts in the landscape. As a clear example of the geographical construction of social fates, small traders who revert to barter tend to be those that do not have access to Chinese business networks. Rahul and Rina's geographical scopes are wider, and their commodities are bought and sold much faster than during their fathers' generation. Such spatial and temporal differences not only divide generations, but also create further stratification between what is considered a "big" trader and a "small" trader, belying an argument the Chinese government often uses to promote economic development in Tibet: that poorer or less-successful traders will eventually "catch up" with the more successful ones if provided with enough modern provisions, tax breaks, or capital stimulus.

These divisions have been seen throughout the narratives I have discussed here and in earlier chapters: while Ramesh travels to buy household appliances in warehouses in Guangzhou and plans to buy land near Nathu-la, Jigme is held still, unable to go across the border to Tibet to bring back packages of Chinese-made toilet paper because of the new ID laws. Unlike Rahul and Rina, who bring foreign products from Vietnam and Australia to Nepal, Gyan and Jeevan find that they cannot sell Nepali products to China anymore and have had to stop and restart their trading numerous times. Although the increasing development of Chinese- and Indian-led infrastructural changes over the past sixty years or so has shifted power away from formerly important trading regions along the Lhasa–Kalimpong route, the processes of reconnecting "lost" places can also occur through seemingly mundane, everyday practices such as bringing a rice cooker over the border to a family member or bartering a scarf for a bowl.

The opening of the passes and the infrastructure projects being built across official borders paradoxically reassert China, India, and Nepal as bounded states. In 2006, those traders who began to sell Chinese goods or those who acquired Chinese or official connections seemed to be doing relatively well compared to those who sold Nepali or Tibetan goods. The latter have found that they must either switch to trading in Chinese goods or find a way to make relevant again certain places that have been cut off. The balance of relative economic power along the Lhasa–Kalimpong route has shifted once more, and traders in the region are feeling the contradictions of capital expansion. As China forges new openings to increase economic development and the "freer" flow of trade, Tibet continues to be drawn more tightly into Chinese economic networks. At the same time, many traders in Nepal and India find themselves cut off from these networks, especially those who do not have relevant linguistic or social access. As a response to this wave of Chinese goods through newly opened spaces, traders reassert and refix their local geographies against new hegemonic geographies in varied ways. We have seen several instances, including the insistence on making the Wool Route stand as historically significant against the rhetoric of the Silk Road, and the reconstituting of face-to-face interactions and immediate exchange—although "by force"—through barter.

That globalization processes produce inequality and are experienced differentially is not news. In the face of hegemonic, large-scale economic changes, peoples' daily practices contribute to the shaping of regional geographies. What I hope to have shown is that there is a general historical pattern—from the micro scale of the commodity to the macro scale of the region—of tensions and contradictions between fixity and mobility, and that through local practices of circumventing state power, people make different kinds of places that are often neglected by state apparatuses. It is these complex tensions between moving and fixing, stopping and starting, opening and closing, that constitute the changing economic and social geographies of the region. Although many of the narratives here assert that the movement of capital in China is driving recent shifts in power and, by doing so, is reasserting state control over its territory and boundaries, we should not think of places (Kalimpong for instance) as merely "victims on the receiving end of globalization" (Massey 2002) or "the local" manipulated by "the global." Instead, places are produced as part of diverse trajectories that are not always part of a linear progression toward neoliberal market triumphalism (Hart 1998). Places, as we have seen, are actively and dynamically created and re-created by the individuals involved in their future.

CHAPTER SIX

Mobility and Fixity

... What does the body remember at
dusk? That the palms of the hands are a map
of the world, erased and drawn again and

Again, then covered with rivers and earth.
Susan Stewart, "The Map of the World Confused with Its Territory"

To Cross a Pass

I heard the phrase "Delhi doesn't understand" yet again from a Sikkimese ex-trader-turned-teacher toward the end of my stay in India. After a long conversation that focused mostly on teaching and education in the region, we moved on to discuss a story he had heard in 2002 about a refugee mother and two children from Tibet who tried to escape to India through the Chorten Nyima mountain pass on the northwest border of Sikkim and Tibet. Apparently, a call went out to the office in Delhi that deals with refugee issues to find out what to do. The authorities in Delhi told the local border guards to send the family back over the pass, and according to the ex-trader, the authorities in Delhi "were not understanding how dangerous the journey was and how tired they were." On the way back to Tibet, the mother and one of the children died.

The ex-trader put his cup of tea down and said very seriously, "Delhi doesn't understand what it's like to cross a pass. We have to get these guys in Delhi to understand what it means to cross a pass. Yaks don't know borders, they cross at will, and when they are restrained, it becomes difficult. Delhi doesn't understand this whole region, and it doesn't understand Sikkim. It doesn't understand the extent of trade that will happen in this region. Border trade has failed everywhere else in India. Someone in Delhi simply thinks, 'OK, we'll open the old border with Tibet and they'll trade salt for grain,' not realizing the implications, not realizing the extent of trade in the past and the potential for today."

As the ex-trader remarked, "what it actually means to cross a pass"—in this case, pure survival—angrily bumps up against what he thinks the state (as represented by Delhi) deems more important: the pageant of border control, the

Sino-Indian conflict, the political sensitivity of allowing or not allowing Tibetan refugees across the borders, and state-based understandings of the realities of border trade in the region. What Owen Lattimore understood all too well in the mid-twentieth century was that routes and frontiers "are shaped less by geographical conditions than by cultural momentum and the impact of those who created them" (Lattimore 1962: 384). While many narratives of traders who have exchanged goods along routes that cross Tibet, India, and Nepal have been influenced and altered by major economic transformations such as border closings and reopenings or the introduction of new commodities, they in turn shape their own trading geographies, often in uneven and unexpected ways. By paying particular attention to competing ideas of place (actual and metaphorical), the processes and struggles over making or "fixing" certain places against others are just as important as the narratives of mobility in globalization studies and other fields, if not more so.

Lived, everyday experiences and the friction of terrain play an important role in the creation and various representations of routes of trade, whether at the Nathu-la border crossing in the early twentieth century and again at the 2006 reopening, or through the contemporary ordering of shipments of French brie to a supermarket in Kathmandu. The economic and political history of the Himalayan region has always been contingent upon tensions between mobility and fixity, at every scale. The stories of the traders who have exchanged goods along the Lhasa–Kalimpong trade route are simultaneously the stories of the geographical development of Asia.

But perhaps it is unsatisfactory to leave our stories at this juncture. I sit here writing this in 2011, and during the past few years, much has been made of the economic rise of China and (to a slightly lesser extent) India. While India has implemented a policy of "Look[ing] East" toward China, China has been putting its "Develop the West" policy into practice since the early years of the twenty-first century. In 2008, China officially became India's largest partner in trade, prompting policy makers to coin the term "Chindia" to reflect the tightening relationship between the two "rising Asian giants." Now with the apparent decline of the US empire and with much of the world in the midst of—or at least anticipating—a deep economic recession, how might we make sense of these tiny slices of Himalayan trading narratives amid the larger story of global historical economic shifts? How might this kind of work lead to future steps in the theorizing of more-just globalization processes? In this final chapter, I point to some further implications in the study of large-scale economic shifts by addressing three related themes that   have emerged thus far, themes that in many measures sum up the larger significance of these stories of Himalayan trade. First, I reexamine the practice of creating geographical diversions in light of Lefebvre's notion of "blind fields" and question how we might conceptualize blind fields in relation to the history of the Lhasa–Kalimpong trade route and to globalization processes in general. Second, I emphasize the historical resilience of trading practices in the Himalayas despite border closings and seemingly potent shifts in modes of production. Third, and finally, since spatial expressions of fixity have been relatively neglected by scholars of globalization in favor of tropes of mobility, I once again suggest that closer attention be paid to how practices "fix" or make visible certain routes or regions against others. These practices are inherent to the survival of capitalism as well as to struggles against dispossession.

Diversions and Blind Fields

In Tibet, the landscape is changing fast, and visibly so. On one crisp winter's day, I took a trip to Gyantse with some friends to visit a small shop where older men and women were said to still be producing colorful hand-stitched Tibetan shoes for traditional dance performances. To get to the shop, we crammed into a taxi that already had two people in it, in order to save some money. We drove past plateau land smattered with buildings and shops nearly all in the middle of being built, mostly Tibetan-style homes that were made with shinier-looking materials than the older, traditional sloped buildings with whitewashed mud and cement walls. I sat in the back with two of my friends, one of whom had a very expensive dark green satin jacket with a real fur collar and a mobile telephone that was covered in small pink rhinestones. She had to half-sit on my lap in order for all of us to fit. Crushed next to her was a loud man, probably in his late twenties, dressed in a tattered sport jacket and holding a cigarette with a dangerously precarious tail of ash dangling at the end. He gave off the stench of alcohol while asking us questions very loudly. Sitting in the front of the taxi next to the driver was a polite young man in an army jacket, who promptly fell asleep as the taxi drove off. The man in the sport jacket began joking lewdly with one of my friends, who subtly made fun of him and rolled her eyes at us while answering.

The man then asked where I was from, loudly asking this question first in my friend's ear, and then in my ear, and my friends decided to tell him—shouting back in his ear while he grimaced—that I was from Dram (on the Tibetan side of the border with Nepal), and in fact spoke more Nepali than Tibetan. "Dram! Near Nepal!" he said. Talk in the taxi turned to that of the newly opened Nathu-la near the Tibetan town of Yatung, much closer to where we were driving. He turned to me. "Did you know that Yadong [Yatung] is now open?" the man shouted. "Did you know that you can go eighty kilometers to the market and can stay for three hours there, and then return? It's not good for Dram, is it?" I agreed. No, it was not good for the trade in the border towns near Nepal, if the roads heading toward Yatung were really developing as fast as they seemed.

The new road to Shigatse had recently been completed, and everyone in the taxi happily discussed how it was so much quicker now to travel on the new diversion from Lhasa to Shigatse—several hours quicker, in fact. There was no more having to go around Yamdrok-tso (a large lake about a hundred kilometers south of Lhasa), and what was even more amazing was how there were fewer accidents when you could drive faster on a smoother road. Talk then turned to the other new road that cut travel time from Lhasa Gonggar Airport to the city of Lhasa by thirty minutes, the comfy seats in the new public buses, and the modern-looking rearview mirrors that made the new buses look like giant insects. "You see them all over China now," the man said, getting a bit quieter. They are a far cry from the buses that were the norm in the late 1990s and early years of the twenty-first century, the rickety, rusty vehicles where you had to fasten all your loads on top. We laughed about how the drivers would always have to stop on the side of the road as someone watched their luggage fall from the roof rack onto the pavement, or how the buses would break down often for some reason or other. My friend in the green jacket was particularly excited about all the new roads, as she was considering purchasing a car once she saved up enough money. Afterward, she would learn how to drive. The man in the sport jacket snorted and asked if we knew the saying, "What are the three most dangerous vehicles in Tibet?" "The answer is this!" he shouted once again. "Chinese soldiers in army trucks, Khampas on motorbikes, and monks on bicycles. But there is a fourth one now. Women in personal cars!"

Things are indeed shifting economically and geographically in this rapidly developing part of Asia. But how do we characterize these processes as they unfold, and how do we talk about what is happening in relation to global capitalism, both ideologically and "on the ground," so to speak? For Henri Lefebvre, the emergence of a "blind field" is the result of an ideological inadequacy that occurs during transitions between large-scale geo-economic shifts. Writing from the particular historical context of the 1968 revolt in Paris, Lefebvre voiced his frustrations with the French Communist Party, specifically their tendency to ignore the urban as a new spatial form and to see it instead as a mere superstructural adjunct above the relations of production (Smith 2003: x). Because the urban is seen only "with eyes, with concepts, that were shaped by the practices and theories of industrialization . . . *reductive* of the emerging reality," according to Lefebvre, party members did not "see" the actual transitions and discontinuities in between industrialization and urbanization (Lefebvre 2003: 29).

Lefebvre's notion of the blind field applies well to the recent history of trade between Lhasa and Kalimpong in similar ways, both geographically and ideologically, for the transitional stages of capitalist processes can sometimes get overlooked in scholarly or media treatments of change in the region. What I mean is this: the lack of scholarly attention paid to "geographical blind fields" is primarily a result of not adequately distinguishing between change driven by capital and change driven by geopolitics.[1] First, the emphasis on state-based development projects and talk of the rise of China and India—as if they were self-contained nation-states with equally contained economies—turns certain trading places and practices into blind fields, or perhaps more accurately, blind spots. In other words, the intensification of Chinese and Indian (and to a lesser extent, Nepali) infrastructural connections highlight state power and "national economies" while obfuscating certain kinds of trading geographies located between and across the margins of states. As we have seen, the focus on connecting the "ancient civilizations" of China and India in relation to the reopening of Nathu-la overlooks bypassed towns such as Kalimpong, as well as specific trading practices such as barter or bringing rice cookers across the border in circumvention of trade restrictions. Seeing globalizing tendencies through the lens of the state economies of China and India obscures other complex trading places and practices that are produced in and across state borders.

Second, much of the rhetoric of contemporary globalization processes concerns mobility, the shrinking of worlds, the erasure of borders, and the increase in transnational and global trade. But my observations seem to uncover quite the opposite. Through these narratives, we see what we have been blinded to: processes that involve making local or regional places more coherent, as well as the kinds of diversions that traders make—sometimes voluntarily, sometimes not—on the turning point of major infrastructural transformations. In the transition to a China-centered economy premised on overcoming territorial barriers for access to new markets, fixing practices have unfolded at the largest of scales: the laying of rail ties to extend the Qinghai–Tibet railway for example, or the implementation of trade restrictions on the Indian side of Nathu-la, or even the introduction of trucks in the 1950s, which displaced transporting separate automobile parts over the passes on yaks and mules. All of these practices reiterated the coherence of national spaces and the power of those who profit from such coherence, even as they simultaneously facilitated the mobility of capital. Small-scale traders also employ place-fixing practices or narratives in order to reap profits, but these practices—such as making visible the different geography of the Wool Route in contrast to the Silk Road—often work against the more hegemonic kinds of practices enumerated above. I continue to elaborate on this notion of ideological blind fields in talking about the resilience of trade below.

The Resilience of Trade

"There is a Chinatown in Kathmandu, why not a Nepaltown in Lhasa?" asked a Newar man who opened his shop in the middle of Lhasa in the 1990s, selling flour, noodles, sweets, rice, coffee, handicrafts, and other items. He claimed that Tibetan customers frequented his shop, and since around 2000, migrant Han Chinese and domestic tourists had started purchasing his items as well. "Tibetans really like Newars because we are religious like them," he claimed. "We went to monasteries in the past and are still religious today." Referring to the long history of Newar trade in Lhasa, he mentioned that "if you were honest and had good relationships a hundred years ago, then a hundred years later, you will still have that reputation." His trading relationship with Lhasans is not new; rather, he has seen quite a few economic transitions throughout his lifetime. His parents and grandparents were members of one of the older Newar Lhasa trading families, and he conducted barter through the governmental foreign trade office in Lhasa in 1986. In 1987, he began using US dollars to trade. "If we wanted to buy wool," he remarked:

> US dollars would be sent from Hong Kong to Tibet, then we would bring the wool to Kathmandu. Before, the Tibetans didn't know how to use the LC [Letter of Credit], but I taught them how to do this two years ago. But let me give you one example of one of the new products I want to export to Kathmandu now. Just this last month [late 2006], I went north to Qingdao, near Beijing, and ordered doors. These doors are made of paper on both sides, with some thin wood inside. I want to make a factory that produces these doors in Lhasa. Wooden doors are so expensive, but these ones are cheap and light. And people want home furnishings for their new homes. There's no chance of the doors shrinking or getting crooked. You see, if you have ideas, so much can be done. The lifestyle is changing in Tibet and Nepal, people want different things.

There is a palpable sense among all traders—large and small—that as lifestyles are changing, so are attitudes toward money. There is a feeling among smaller traders, however, that those who are becoming more and more involved in Chinese business networks have been blinded to the reality of the struggles of the less-fortunate traders. According to Fatima, the daughter of the jewelry trader in Kathmandu introduced in chapter 2, "Tibetans in Lhasa have a different mindset from us [Tibetans living in Kathmandu]. They have changed. Now they have a Communist mentality because they only think about money all the time." Fatima made this somewhat ironic observation in English during a conversation that revolved around the differences between her family in Nepal and some of her relatives still living in Lhasa. One could, of course, read the comment about "a Communist mentality" in many ways: a translation mix-up was the first thing that came to mind. But I couldn't shake off the fact that I heard similar statements from others throughout my fieldwork. While sitting in an elderly Chinese Muslim man's house in Lhasa overlooking the vegetable vendors in one of the back alleyways in the older section of town, my friend recalled what education was like in China during the time of the Kuomintang. It was "helpful," he said, because Muslims were offered free accommodation, clothing, and education at mosques. In the 1940s, he said, the "Chinese Capitalism Party" was just beginning to form, and as it was very poor, there was not enough money to provide resources such as free education. Even though I asked—twice—if he actually meant the Chinese Communist Party, he kept referring to the CCP as the "Chinese Capitalism Party" without any irony. Without wishing to make too much of these stories, I believe that the linguistic entanglement of capitalism and communism is actually reflective of traders' lived experiences of transitional geographical and economic shifts, as well as what China really is at the moment. These kinds of shifts are not always visible when it comes to theorizing geopolitical transformations.

It is tempting to argue that the history of trade in Tibet follows a trajectory based on distinct shifts in modes of production. In other words, it would appear on a *very basic level* that pre-1950s Tibet had some amalgamation of a severe socioeconomic hierarchy concentrated in the hands of the elite and the monastic estates (though the CCP would argue "feudal"), then it became Communist ("liberated"), and after the 1978 reforms it became subject to a postsocialist ("socialism with Chinese characteristics") plan to develop and securitize the "backward" western areas of the nation. But, as the story here suggests, such simplified portrayals of macro political-economic shifts do not accurately represent the reality of what happens on the level of trading practices, where various modes of production overlap and are in tension with each other. At a time when markets are increasingly taking precedence over territory, it is imperative to avoid conceptualizing capitalism as something that is uniquely "owned" or generated by the West, and to take the capitalist *processes* that are currently occurring in and throughout Asia more seriously, not as some special variety or separate strand of "Asian" or "Chinese" or "Indian" capitalism—or even, as Giovanni Arrighi provocatively contends, a distinct kind of Chinese noncapitalist market society (Arrighi 2007)—but as part and parcel of global hegemonic capitalist processes in general (Cammack 2010).

Shifts in large-scale economic geographies have less to do with any abstract changes in modes of production than they do with concrete geopolitical and economic events—in this case, border closings and openings, the introduction of new commodities and materials such as makeup sets in the 1940s or silk apron thread from Chengdu in the twenty-first century, and the emergence of new kinds of infrastructure and modes of transport, exemplified by the shift from mules to trucks. There are thus never any clean temporal ruptures between modes of production: for instance, although writers often earmark 1951 as the year that Tibet "became Communist", there was a period when huge private profits were to be made by elite Tibetan and Indian traders selling Indian goods to the new influx of PLA soldiers in Lhasa. Moreover, to take a current example, poorer, small-scale traders do not stop attempting to trade when faced with barriers to accessing profitable Chinese trading networks or the supposed "free flows" of globalized international trade. Instead, they survive by diverting in various ways and by resorting to barter and other forms of capital exchange. Regardless of restrictions, and despite apparent major shifts in modes of production since the 1940s, trade in this region continued to evolve. So what may be seen on a macro level as a historical shift from some amalgam of feudalism to communism is not so seamless. What is important for scholars of globalization to examine, then, is individuals' lived experiences, responses to, and transformations of major spatial changes (such as border openings or closings), not necessarily how sociohistorical shifts are tied to specific modes of production.

Uneven and stratified trading experiences are very much a result of the Chinese government (and to a lesser extent Indian and Nepali governments) implementing intensified economic policies and development projects to open up borders for increased trade and national security. Even though most traders do not expect trade at Nathu-la to pull in massive revenue over the next few years, long-term plans such as eventually opening the route to tourists as well as encouraging investment in factories near Yatung indicate that this border region is earmarked as one of the many focal points for continued future development. Despite infrastructural fragility on the Indian side of Nathu-la (a serious worry due to environmental hazards in the hilly, monsoon-prone area), there are now several additional openings for Chinese goods to be moved through other South Asian, Southeast Asian, and Middle Eastern borderlands, as outlined earlier. The future proposal—whether true or not—to extend the Qinghai–Tibet railroad up to Nepal and to Nyingchi on the Assam border is just one example of these many "opening" mechanisms that are simultaneously fixing mechanisms. In addition, and hand in hand with the accelerated opening of border areas and promotion of western Chinese development, comes a 15 billion–yuan project expected to last until 2030, literally translated as the "construction of Tibetan Plateau national ecological security barrier." This government-run project proposes to set aside *Tibet itself* for environmental conservation. It includes the protection of the grasslands, the construction of nature reserves, the cultivation of seeded grasslands, the protection of water and soil erosion, and the recovery of abandoned mines. Fixing Tibet itself as a "security barrier" in the name of environmental protection in fact sets it up as a springboard for further opening and development plans: this spatial fix creates more opportunities for opening and the flow of capital.

Perhaps it is time to return again to Owen Lattimore's work on inner Asian frontiers and to take it more seriously, beyond merely an area-studies audience. In the epigraph to the previous chapter, Lattimore commented in 1953 that China was "making a turn away from the coast and in towards the center" and that it was "expected that the Chinese will go a long way in moving populations to deep hinterlands to build industrial centers . . . " (Lattimore 1962: 176). This was indeed a brief military strategy during the Third Front period (1964–1971), when China began to build heavy industrial infrastructure in the Tibetan areas of Kham and Amdo (Sichuan, Qinghai, and Gansu) as a line of defense against Soviet, Indian, and US invaders from the coast, but that strategy ended with the US détente in 1971–72 (Fischer 2005: 18). Today, Chinese state and foreign investment is building up the borderlands of Asia again: urban Lhasa is officially supposed to have expanded from three square kilometers to over fifty-three square kilometers in a period of fifty years, factories are being built in Yatung, and by late 2008 over 21 billion yuan (US$3 billion) was earmarked for mining, construction enterprises, and food processing factories in the TAR (Xinhua 2008).[2]

As Tibet turns into a regional center of inner Asia, it becomes at the same time more and more linked to the Chinese nation-state. The opening of borders for trade does not so much facilitate the "free flow" of goods as it reiterates—fixes—the coherence and the power of the state. Furthermore, while borders and bordering practices are increasingly unattached to specific territories, they still "strengthen state space as a 'bounded unit'" (Johnson et al. 2011: 62–63). This strengthening of the state's power remains stubbornly resilient throughout long-term history, becoming more entangled when people ostensibly "from below" cooperate with the state to the extent that it is advantageous to them. Much of what seems to be happening on the national borders of China echoes what Gerry Kearns has said regarding the United States' expansionist policies in the early years of the twenty-first century: that they "will intervene in places where the inhabitants do not have the market-given right to explore (commercial) opportunities. The US will police a world in which borders are about markets and not territory" (Kearns 2006: 73, 87). Though it was Britain that "opened" Tibet through India in the early twentieth century in order to increase imperial trade as well as to stave off potential Russian expansion, diversions and tensions between mobility and fixity are just as historically contingent today. In order for China to move commodities to new markets, it must overcome territorial obstacles such as mountains and open its national borders; at the same time, however, these kinds of moves are supposed to maintain national security and quell dissent in the "unstable" western regions.

Mobility and Fixity: Diversions and New Directions

When I first began writing about these narratives of traders in the Himalayas, I sent half-formed chapters to peers who were willing to read and comment on them. One of my colleagues in my graduate anthropology department (who himself is from Nagaland, one of the northeastern states of India, and very familiar with the landscape of the region) confessed that he felt physically exhausted after reading my chapters. He said that he imagined himself traveling in a shared jeep from Darjeeling to Kalimpong, then perhaps taking a train to Siliguri, and then taking a plane from Kathmandu to Lhasa. All of this "traveling"—through checkpoints and across monsoon-covered roads, waiting to board a plane—made him tired. At first I was concerned, thinking that it really wasn't fair to the reader to have to decipher a piece of writing so unsettled, shifting, and peripatetic. It is indeed jarring to read a book that moves back and forth across three different countries, with snippets of narratives from people in multiple towns during different historical periods. In retrospect, though, I think it may actually be necessary for research on evolving globalization processes to be a little geographically exhausting.

Let me try to defend myself here. The movement of people or goods from Kalimpong to Lhasa—or from Kathmandu to Guangzhou, for that matter—has never been totally smooth and without incident. Instead of crafting an unbroken, fluid story of trade in the Himalayan region, I have explored various disjointed "snapshots" of trading practices and policies ("ethnographic fragments," as Anna Tsing has called them), not just "from below" or by "studying up," but on several spatial levels and for various temporal periods. These snapshots, while unavoidably selective, highlight the jostling experiences of moving and fixing, of having to pick up and relocate if a border is closed, or of having to switch and divert to a more lucrative trading strategy in a different location. Even in the context of this research in 2006, when globalization processes were commonly characterized in the popular media by increased movement and fluidity, the emergent fixities in this region were extremely evident. Economic development projects in these border regions became sites for struggles over making certain kinds of places coherent against others. The reopening of Nathu-la prompted the emergence of groups for and against the reopening of Jelep-la, the marketing of non-Tibetan goods as "from Tibet," and the resilience of barter against barriers to trade—all practices that are often "unseen" if we focus too much on tropes of global mobility.

Why am I not framing this argument in terms of paying attention to the creation of "alternative geographies" of globalization, then? Why is looking at the tensions between fixity and mobility on multiple scales a better model for examining long-term globalization processes? In my opinion, the notion of "alternative geographies" of globalization—that is, various kinds of place-making practices with motives and goals that differ from hegemonic capitalist ones—tends to elide the fact that practices of trade are not always "alternatives." In fact, as we have seen, traders do not always work directly against hegemonic geographies; at times, they work to reinforce them. Take, for example, traders who have named their Kalimpong-based shop "Silk Road Trading Company" even though they believe it was always located on the "Wool Route," or those who have marketed items produced in Guangzhou or Nepal as "genuinely Tibetan" in order to reap the benefits of Chinese development and gain a monopoly price. Even further, what about Lobsang's refusal to draw the "alternative" trading map I requested at the very beginning of my research for this book? A glimpse at these unusual and often paradoxical tensions between practices of mobility and fixity moves us beyond reducing global processes to modes of production and beyond the idea that "solidarity means homogeneity" or "hegemony means capitalism" (Tsing 2005: 245). Taking into account the surprisingly awkward bedfellows and "unexpected collaborations" that are sometimes produced in struggles over making places coherent raises the need for a better understanding of the very contradictory experiences of long-term, global economic shifts.

On a disciplinary note, if anthropology has been typically characterized by an ethnographic methodology that involves examining culture on a small scale via long-term fieldwork, and if geography is seen to focus on larger-scale human interactions with place and the environment, the study of cross-border experiences of globalization necessarily demands that we forge connections between traditionally bounded disciplines and borrow from other methodologies and theoretical canons. Perhaps this work can begin by rethinking mobility and capitalist processes. In the sociohistorical evolution of trade in Asia, fixity and mobility have always gone hand in hand at every scale, but we seem to be blinded to the intense struggles over fixing place in moments of economic transition. The building of structures meant to facilitate trade, such as new roads in Sikkim, and the introduction of new modes of transportation (be they yaks, railroads, cars, or trucks) are also sites where opposing groups struggle to make their own trading places coherent or visible. There is therefore a considerable need to go beyond simply presenting the fact that contradictions or tensions or paradoxes *exist* at borders. Greater attention needs to be paid to individuals' complex experiences of spatial transformations such as border openings or closings, asking how they in turn create diversions by fixing or making their own trading places in answer to these major geographical shifts.

A FURTHER NOTE ON RESEARCH METHODS

My research methodology warrants a bit more explanation than has been given in the body of this book. Here, I briefly touch on some issues regarding surveillance, research permits and visas, interview languages, and the specter of illicitness and smuggling. Anthropologists tend to spend a significant amount of time living with local families in order to gain cultural insights that might not otherwise result from sporadic interview sessions. In Lhasa, however, since foreigners are forbidden to stay overnight in places (such as certain hotels and all private residences) that are not officially authorized to host foreigners, living with a Tibetan family would have required a special permit from the Public Security Bureau. Because of this restriction, my partner at the time (who was conducting textual research in the TAR) and I ended up living for six months at a hotel in the old section of Lhasa, and I would venture out from there each day to conduct interviews. Although this arrangement was not ideal, it did allow for us to live centrally and provided us with some privacy.

In a heavily surveilled city like Lhasa, a sense of paranoia is not uncommon, as other researchers who have worked in the city, such as Emily Yeh (2003; 2006) and Robert Barnett (2006), have attested. Surveillance techniques such as keystroke-logging software programs and plainclothes "minders" are part of everyday life for many people in the city. Because there is no way to know if and when any of these methods is actually being employed, it is more often the feeling of *not* knowing that fuels paranoia. For instance, one day in Lhasa, a young monk approached me at the gate to my hotel and asked me several times for a *par* (photograph). I assumed he meant a photograph of the Dalai Lama, which is illegal to possess in the TAR. Since most people in Lhasa—especially monks and nuns—are aware of this restriction, and I knew that the gate of the hotel was constantly watched, I speculated that he must have been a spy of some sort. As the monk kept pressing me for a photograph, I pretended that I didn't understand him very well and thought that he wanted me to take a photograph of him. I made camera-clicking gestures, and said, "A! *par rgyag! par rgyag!*" (*par rgyag* means "to take a photo"). As I took out my camera to take a photo of him, he quickly walked away from the gate, in full view of the watching guard. In the end, however, I had no way of telling whether my assumption was correct, if my actions were simply foolish, or if I failed some sort of "test." This kind of worry permeated a lot of my fieldwork in Tibet.

These heightened emotions become more palpable in my field notes toward the middle of my stay in Lhasa:

> How it makes me so angry to think that people might be listening to everything I write, do, say, breathe. Am fuming, also because I can't get deep answers to questions—I just get the same rote answers, like a survey—people probably want to talk, but can't. So frustrating. I can't ask, I can't "go there," I can't even ask about [trade] routes because now I am scared. Shit. And this is all made worse by yesterday, when a boy in a wheelchair with a dog told the dog to bite me, and his friend laughed.
>   . . . I then went to the tailor shop, but it was closed (on XX Street), then the coral/turquoise place, where the owners were nice, but the shop was very crowded. It then dawned on me that there were people following me again, and now I just have to throw my hands up—like with the Mr. Z situation [a "minder," hired as a "chef" during another trip to Tibet, who never cooked a thing in our presence during the three weeks we were in Tibet], to just joke about it, and to know that everyone knows I know (though I don't really).

There are strict official regulations for conducting academic research in India and China. In the TAR, I was kindly given official research permission and visa clearance through the Tibet Academy of Social Sciences and the Tibet and Himalayan Digital Library (now called THL) to conduct research on material culture and old trading practices in Lhasa, with permission for a few side trips to other major cities in Tibet. In India, I received a research visa through the Ministry of Education of the Government of India and arranged affiliation with the International Trust for Traditional Medicine in Kalimpong. In Kathmandu, I was able to obtain permission for research through Tribhuvan University. Although I would have liked to have traveled along much of the length of the trade routes, it is nearly impossible to conduct research near any national borders for obvious reasons: they are extremely sensitive, militarized zones. My methodology therefore relied on staying within the cities of Lhasa, Kalimpong, and Kathmandu, chatting with elderly traders who recalled their memories of pre-1962 trade, and interviewing younger traders who often came back from border trading on the very day of the interview.

In Kalimpong and Kathmandu, I mostly used English with a smattering of Nepali during interviews, although a few times I needed to rely on the translation skills of the relatives of elderly traders if their first languages were Newari or Marwari. In Lhasa, my Tibetan was adequate enough initially to conduct simple interviews and improved quite rapidly over the course of the fieldwork period. Perhaps the biggest drawback to my research in the TAR, however, was my inability to speak adequate Chinese. Although I could successfully buy eggs or pork buns at the grocery store or bargain with bicycle rickshaw drivers, I was not in a position to interview anyone on my own. While in Lhasa, I specifically asked if I could locate someone who could assist me in case I needed to interview Chinese-speaking shopkeepers. The translator I found turned out to be absolutely invaluable as a collaborator, not just during the few Chinese-language interviews I needed to conduct, but also during several note-taking sessions after Tibetan-language interviews. As many individuals in the TAR understandably refused to be tape-recorded, I would conduct interviews by taking brief notes and then typing them up as soon as possible. This method of collaborative note taking was very useful, as there were times she would remember crucial parts of the conversation that I didn't, and vice versa. I never brought up politics in my interviews in the TAR, and in fact, some of the braver interviewees would ask me bluntly at the beginning of our conversation, "This isn't about politics, is it?" Open-ended interviews included questions such as "What were some of the commodities that you traded?" and "How have things changed since your parents' generation?" But occasionally, attitudes toward Chinese governance in the TAR would seep through some of the answers, and I would often feel uncomfortable during these moments.

Trade, especially cross-border trade, is often associated negatively with smuggling, prostitution, drugs, and general illicitness. One of the questions I have been frequently asked is whether I encountered any evidence of drugs, arms, or human trafficking across borders. As exemplified by the story of Jigme and his trading experiences, Carolyn Nordstrom has accurately shown that even though one would suspect that drugs travel along a "drug route" and arms along a separate "arms route," they in fact get mixed up and travel along the same channels. Although I heard plenty of low-level smuggling stories, I actively avoided asking questions about such things. However, I did find myself in a couple of sticky situations at the beginning of my fieldwork. As soon as friends knew I was interested in studying trade, many people would offer to take me to see all kinds of traders. I accepted most invitations, visiting cheese traders, wool traders, medicine traders, and cosmetics traders, who would of course not always stick to selling those single categories of items. In order to illustrate how the legal and extra-legal are entangled in trade, here is an excerpt from my field notes from a visit to a couple of Tibetan semiprecious stone traders on my own:

> Then the guy reaches into a cupboard and pulls out a small wooden statue, about the size of my palm, which fits into a wooden reliquary box, and then, whispering, he pulls out a huge *tsa tsa* [clay votive objects that are created as part of

meditation practice], which is wrapped in a pink towel and placed in a China Mobile box—both of these, apparently, are supposed to be from X [the name of an old monastery]. Then he makes motions like he is hiding the wooden box in his pocket, and says that it was difficult to do the same with the *tsa tsa*. I am visibly worried, and Y says he has nothing to do with this, he just buys them when they are on the market—but monks sometimes steal them and sell them. Aiee. As we sit surrounded by piles of things from Bhutan, Afghanistan, the far reaches of China, Mongolia, and Tibet, I have no idea how all these things got here, what things are real, what are fake, what are stolen, and what are legitimately family heirlooms. Rather boggling. I begin to get scared that there are gangsters looking at us through the windows, demanding money or compensation for something.

After another similar incident, I stopped accepting invitations to visit *all* traders and tried to concentrate on the seemingly more benign wool trade. I truly appreciate the generous help by the traders who knew I was interested in trade and took me around to various merchants, but I could not continue witnessing and then turning my back on practices involving items that were likely stolen from monasteries, compromising both the national laws on antiquities and social scientific codes of ethics.

Since I lack adequate formal Chinese-language abilities, I was unable to consult important documents such as historical Chinese gazetteers. I am also aware that there is a significant and noticeable gap in the data concerning Tibetan and Chinese trade and politics between the 1960s and 1980s, particularly during the Cultural Revolution. I mention these deficiencies because I know that I have just scratched the surface on contemporary trade in the Himalayas, and it is my hope that the next generation of researchers (especially those who speak and read both Chinese and Tibetan fluently) can provide a corrective and add significantly to these initial findings.

NOTES

INTRODUCTION. *Tibet, Trade, and Territory*

1. Because of the sensitive nature of conducting research on cross-border trade and in Tibet, an area with potential political, economic, religious, and cultural unrest, I have changed the names and many identifying characteristics of people and several locations in this book. Although some individuals said that they would not mind my using their real names, I have given everyone pseudonyms for the sake of consistency. See both the methodology discussion at the end of this chapter and the note on research methods at the end of the book for further information on fieldwork methodology and language.

2. Don Mitchell, citing Raymond Williams, points out that one of the purposes of creating a landscape is to make the view look natural, as if it were made on its own, for "landscape is both a work and an erasure of work" (Mitchell 1996: 6).

3. The city of Calcutta was renamed Kolkata in 2001, partly in order to deemphasize British influence and to highlight its Bengali identity. I will refer to the city as Kolkata throughout the book, both for the sake of consistency and also because the historical naming and renaming of places is another example of configuring certain kinds of territory against others.

4. For a good description of British economic interests in Tibet, see Schuyler Cammann, *Trade through the Himalayas: The Early British Attempts to Open Tibet* (Westport, Conn.: Greenwood Press, 1970).

5. In 2001, the population of Kalimpong was approximately 43,000 inhabitants. (At the time of writing, the figures for the 2011 Indian census were at a provisional stage and available only at the district level.) The official 2000 census figures for urban Lhasa were roughly 230,000 (residents of "urban administrative areas" in all of Lhasa municipality). See Yeh and Henderson 2008 for more information on TAR census figures. In 2001, the population of Kathmandu was approximately 672,000, and the projected figure for 2010 was nearly 990,000.

6. The mass protests in the TAR in March 2008 and Chinese worries over security before and during the 2008 Olympics have led to the indefinite cancellation of most of these partnerships. During 2009, 2010, and 2011, the TAR has also been closed to foreigners on several separate occasions, for instance in anticipation of possible Tibetan unrest during the sixtieth anniversary of China's rule over Tibet, the fiftieth anniversary of the exile of the Dalai Lama to India in 1959, the anniversary of the March 10, 1959, uprising, or the anniversary of the huge demonstrations that began on March 5, 1989.

CHAPTER ONE. *Middlemen, Marketplaces, and Maps*

1. The history of the *Mirror* is explained in further detail in books such as Tsering Shakya's *Dragon in the Land of Snows* and H. Louis Fader's Christian bio-history of Tharchin Babu; however, nearly the full run of newspapers has now been collected as part of the Tharchin Collection at the Starr East Asian Library, Columbia University.

2. This clipping, taken from the June 28, 1938, issue of the *Tibetan Mirror* newspaper, reads, "A very useful, top quality and strong carding comb for female wool carders. Please check whether or not the 'woman carding wool label' is there when you make your purchase." Isabelle Henrion-Dourcy has noted that the Tibetan word for "label," "trademark," or "brand" (*lam 'bar*) "is held to be a Tibetan phonetic adaptation of either 'number' or 'label' in Indian English," since many foreign products were more easily distinguished by their image (e.g. "peacock brand" soap) than by their written name. The term *lam 'bar* was also used to nickname or "brand" *nang ma* (Tibetan classical music) singers and prostitutes; according to Henrion-Dourcy, *sha bag leb lam 'bar* (meat-filled bread label) was the nickname for a woman whose mother sold such bread, and *sna rtug lam 'bar* (snot label) was someone who was likely often seen with a runny nose (Henrion-Dourcy 2005: 204, n. 22).

3. *Xibu da kaifa*, the "Develop the West" campaign, was formally initiated in 2000, coinciding with plans for China's entry into the World Trade Organization. Since then, over US$150 billion has been allocated to hundreds of economic and political projects in western regions of China (particularly the Autonomous Regions of Tibet and Xinjiang) in order to promote "national unity and state security," as outlined in Hu Jintao's speech in Lhasa, July 19, 2001, on the occasion of the fiftieth anniversary of the Peaceful Liberation of Tibet. These development projects include attracting foreign investment, building infrastructure projects such as the Qinghai–Tibet railroad and the Three Gorges Dam, and encouraging migrant workers to move to western areas.

4. For some fascinating lists and descriptions of Tang Dynasty trade goods, see Edward Schafer's book *The Golden Peaches of Samarkand: A Study of T'ang Exotics* (1985).

5. For an excellent, detailed historical account of the mapping of India, see Matthew Edney's *Mapping an Empire: The Geographical Construction of British India, 1765–1843* (1997).

6. I barely skim the surface of the history of the British in Tibet; for more detailed historical accounts, see Alex McKay's publications, including *Tibet and the British Raj: The Frontier Cadre, 1904–1947* (1997).

7. Carole McGranahan's excellent PhD dissertation (2001) details the rise and fall of the Pangdatsang family.

8. "Old documents in the Kalimpong sub-divisional office show that trade ties between India and Tibet were so close that three Kalimpong-based private banks—Kuber Bank, Das Bank and a third bank that belonged to Rai Bahadur Ramchandra Mintry—operated in Tibet till the early 1950s" (Chaudhari 2003: 18).

9. According to the former traders who spoke with me, the travel time from Lhasa to Kalimpong on mules ranged from fourteen to twenty days in the early 1950s.

10. The Tsarongs were an elite Lhasan family, well known for their influence and involvement in early twentieth-century Tibetan politics.

11. See Kesar Lall, *The Newar Merchants of Lhasa* (2001); Deb Shova Kansakar Hilker, *Syamukapu: The Lhasa Newars of Kalimpong and Kathmandu* (2005); and Kamal Tuladhar, *Caravan to Lhasa: Newar Merchants of Kathmandu in Traditional Tibet* (2004) for a more detailed history of Newar traders in Lhasa.

12. "Baba" is a term of respect in Urdu/Hindi, often used for holy men.

CHAPTER TWO. *From Loom to Machine*

1. Although there is an extensive literature on Tibetan carpets, there is a paucity of in-depth documentation on other kinds of textiles, especially, it seems, *pang gdan* weaving in Tibet. For an excellent, more comprehensive account of textile production and gender in Ladakh, where weaving techniques are similar to those in Tibet, see Monisha Ahmed's *Living Fabric* (2002).

2. Rolls of *snam bu* are measured in outstretched arm's-length measurements (*dom*).

3. Coral was originally was brought into Tibet in the thirteenth century from places as far afield as Italy and the Red Sea (McKay 1997: 54–55); more contemporary sources are Japan and Taiwan.

4. For more information on Tibetan environmentalism, see Emily Yeh's forthcoming paper, "The Rise and Fall of the Green Tibetan," in *Mapping Shangri-la: Nature, Personhood and Sovereignty in the Sino-Tibetan Borderlands*, edited by Emily Yeh and Chris Coggins (based on Yeh 2007a). Following Anna Tsing's work in *Friction* (2005), Yeh traces the development of a Tibetan environmental discourse. Originally based on collaborations between exile Tibetans and westerners, this discourse takes on a very different kind of subjectivity in the TAR as a result of both new Chinese links to environmental NGOs and increases in state control.

5. There are a number of good studies that address the geographical separation between producer and consumer and the strategic marketing of place in order to ostensibly bridge the gap between buyer and seller. Notable examples include Carol Hendrickson's work (1996) on Guatemalan handicrafts sold in western mail-order catalogs and Sharon Hepburn's work (2000) on "Thamelwear," clothing sold in Kathmandu that is marketed as "Tibetan" but actually produced by Nepalis.

CHAPTER THREE. *Silk Roads and Wool Routes*

1. It is worth noting that Kiran Desai's 2007 Booker Prize–winning novel set in Kalimpong, *The Inheritance of Loss*, came under much scrutiny in the region for its negative and condescending portrayal of Nepali-speaking inhabitants in the mostly Bengali-speaking state of West Bengal. One local reaction was that "it is a one-sided account that tells you about [Desai's] fears about Kalimpong. The central character Sai is obviously a self-portrait and you can feel her estrangement from this dark, ominous place where Nepalese are just transient interlopers in the landscape" (Ramesh 2006).

2. "La" means "mountain pass" in Tibetan. In order to avoid redundancy, I henceforth refer to the passes simply as "Nathu-la" or "Jelep-la."

3. Several Tibetans thought it was no coincidence that Nathu-la was reopened on the Dalai Lama's birthday, and wondered if the timing was a deliberate Chinese attempt to undermine his power.

4. Separatist movements, such as the Gorkhaland movement calling for a Nepali state in Sikkim and the North Bengal hills, have marked much of the recent history of the eastern Himalayas. I am fully aware that I do not do this crucial issue much justice in this book.

5. West Bengal held a Communist-led government (Communist Party of India [Marxist]) from 1967 to 2011.

6. The Teesta is the major river that flows through Sikkim and North Bengal. In 2010, the Teesta Low Dam hydroelectric project was at stage 4, wherein 103 hectares of land would be submerged, inevitably contributing to increased erosion and landslides. It was supposed to be completed by autumn 2009, but by July 2011, the imminent flooding of Lepcha tribal territory had given rise to mass hunger strikes and protests by groups such as the Affected Citizens of Teesta. In all of the state of Sikkim, thirty-five new hydroelectric projects have been planned or have been under way since 2001, but up to four of these projects were cancelled in 2012 because of public opposition and safety concerns following the 6.9 magnitude earthquake in Sikkim on September 18, 2011. See McDuie-Ra 2011 for a more detailed discussion of the tensions surrounding hydroelectric projects in Sikkim.

7. As one Marwari salesman in Siliguri said, "Before the agitations, Kalimpong was the hub of wholesale business for Sikkim—at the time it was mostly wholesalers and very few retailers. In 1986 we came to Siliguri, because of the agitations, thinking it would be a long time until things started settling down. I opened my shop here on Sevoke Road on 27 July 1986, the day of the worst agitation in Kalimpong, so my parents couldn't come for the opening." Here, "the agitation" (and very likely K. C.'s mention of "political issues") refers to the Gorkha National Liberation Front's active call for a separate Nepali state in the 1980s and the violent clashes that ensued.

8. During the first years of the twenty-first century, ginger crops in North Bengal, Sikkim, and Bhutan fell prey to a fungus that cut production by nearly half.

CHAPTER FOUR. *Reopenings and Restrictions*

1. Then again, much of the "Roof of the World" imagery—found not just in the West but within China as well—tends to reinforce the idea of Tibet as an inaccessible, mysterious, and remote land.

2. Also see, for example, Abraham and van Schendel 2005; Aggarwal 2004; Andreas 2000; Konstantinov 1996; van Schendel 2005; Wilson and Donnan 1998.

3. Abu-Lughod 1991; Arrighi 1994; Braudel 1982; Castells 1996; Curtin 1984; Harvey 1990; Lattimore 1962; Marx 1976; Polanyi 1944.

4. I take some issue with the term *informal economies* insofar as it sets itself up against and assumes such a thing as a "formal" economy, but as it has been applied by many anthropologists to mean simply "unregulated business activities or services," I will use it as such here. For additional ethnographic accounts, see Abraham and van Schendel 2005; Cheater 1998; Konstantinov 1996; Nordstrom 2007; Stephen 2007.

5. Currently, only residents of Sikkim are allowed to obtain trading permits. This major restriction leaves out other traders from Kalimpong and other towns in nearby West Bengal.

6. Tsering Shakya notes that the eating of *tsampa*—not necessarily religion, language, or geography—is often what connects diverse groups of Tibetans, including those in Ladakh and Bhutan (Shakya 1993).

7. Here and in the subheading to this section, I borrow from Carolyn Nordstrom and Abraham and van Schendel's work. Sectioning off the words (il)licit or il/legal in this way draws attention to the intertwined and often indistinguishable worlds of licit and illicit, legal and illegal activities (Abraham and van Schendel 2005; Nordstrom 2007: xviii).

CHAPTER FIVE. *New Economic Geographies*

1. I thank Emily Yeh for this observation.

2. Actually, roads to Lhasa from Sichuan, Qinghai, Yunnan, and Xinjiang were built in the 1950s, and a road to Nepal was built in the 1960s.

3. As mentioned in chapter 2, a lot of "culturally Tibetan" items in the tourist markets of Tibet are in fact made in Nepal.

4. For a more in-depth analysis of how these ideas of self-deprecation are used in the Tibetan context, see Emily Yeh's article "Tropes of Indolence and the Cultural Politics of Development in Lhasa, Tibet" (2007b).

5. The raw materials for *kha btags* sold in Kalimpong often come from Orissa, but the printing of auspicious symbols, cutting, and separation are done in Kalimpong.

CHAPTER SIX. *Mobility and Fixity*

1. I thank one of my book's peer reviewers for putting it more clearly in these words.

2. The administrative (as opposed to "real") figures for the expansion of urban Lhasa are likely to be scaled up in accordance with the discourse of urbanization-as-progress in many developing areas of China (Yeh and Henderson 2008: 16).

REFERENCES

Abraham, Itty, and Willem van Schendel, eds. 2005. *Illicit Flows and Criminal Things: States, Borders, and the Other Side of Globalization*. Bloomington: Indiana University Press.

Abrams, Phillip. 1988. "Notes on the Difficulty of Studying the State." *Journal of Historical Sociology* 1(1): 58–59.

Abu-Lughod, Janet. 1991. *Before European Hegemony: The World System A.D. 1250–1350*. Oxford: Oxford University Press.

Adams, Robert McC. 1974; repr. 1992. "Anthropological Perspectives on Ancient Trade." *Current Anthropology* 33(1): 141–60.

Adams, Vincanne. 2002. "The Sacred in the Scientific: Ambiguous Practices of Science in Tibetan Medicine." *Cultural Anthropology* 16(4): 542–75.

Aggarwal, Ravina. 2004. *Beyond Lines of Control: Performance and Politics on the Disputed Borders of Ladakh, India*. Durham, NC: Duke University Press.

Ahmed, Monisha. 2002. *Living Fabric: Weaving among the Nomads of Ladakh Himalaya*. Bangkok: Orchid Press.

All China Marketing Research Co. 2008. *Tibet Statistical Yearbook 2008*. Beijing and Ann Arbor: University of Michigan China Data Center.

Amster, Matthew H. 2005. "The Rhetoric of the State: Dependency and Control in a Malaysian-Indonesian Borderland." *Identities: Global Studies in Culture and Power* 12(1): 23–43.

Anand, Dibyesh. 2008. *Geopolitical Exotica: Tibet in Western Imagination*. Minneapolis: University of Minnesota Press.

Anderson, Benedict. 1992. *Long-Distance Nationalism: World Capitalism and the Rise of Identity Politics*. Amsterdam: Centre for Asian Studies Amsterdam.

Andreas, Peter. 2000. *Border Games: Policing the US-Mexico Divide*. Ithaca, NY: Cornell University Press.

Anna, Cara. 2008. "In Remote China, Tibetans Break Silence." Electronic document, http://abcnews.go.com/International/wireStory?id=4489994, accessed March 22, 2008.

Appadurai, Arjun, ed. 1986. *The Social Life of Things: Commodities in Cultural Perspective*. Cambridge: Cambridge University Press.

———. 1991. "Global Ethnoscapes: Notes and Queries for a Transnational Anthropology." In *Recapturing Anthropology: Working in the Present*, ed. Richard Fox, 191–210. Santa Fe: School of American Research Press.

Arora, Vibha. 2008. *Routing the Commodities of the Empire through Sikkim, 1817–1906*. Commodities of Empire Working Paper No. 9. Milton Keynes and London: The Open University and London Metropolitan University.

Arrighi, Giovanni. 1994. *The Long Twentieth Century: Money, Power, and the Origins of Our Times*. New York: Verso.

———. 2007. *Adam Smith in Beijing: Lineages of the Twenty-First Century*. New York: Verso.

Atreya, Sarikah. 2007. "Second Round of Trading at Nathu la—Will It Set the Cash Register Ringing?" Electronic document, http://www.thehindubusinessline.com/2007/04/13/stories/2007041301010900.htm, accessed March 20, 2008.

Aziz, Barbara. 1975. "Tibetan Manuscript Maps of Dingri Valley." *Canadian Cartographer* 12(1): 28–38.

Barnett, Robert. 2006. *Lhasa: Streets with Memories*. New York: Columbia University Press.

Barth, Fredrik. 1988 [1969]. *Ethnic Groups and Boundaries*. Boston: Little, Brown and Company.

Baud, Michiel, and Willem van Schendel. 1997. "Toward a Comparative History of Borderlands." *Journal of World History* 8(2): 211–42.

Bauer, Kenneth. 2004. *High Frontiers: Dolpo and the Changing World of Himalayan Pastoralists*. New York: Columbia University Press.

BBC News Online. 2008. "'Free Tibet' Flags Made in China." Electronic document, April 28, 2008, http://news.bbc.co.uk/2/hi/7370903.stm.

Beckwith, Christopher. 1977. "Tibet and the Early Medieval Florissance in Eurasia: A Preliminary Note on the Economic History of the Tibetan Empire." *Central Asiatic Journal* 21(2): 89–104.

Behdad, Ali. 2005. "Postcolonial Theory and the Predicament of 'Minor Literature.'" In *Minor Transnationalism,* ed. Françoise Lionnet and Shu-mei Shih, 223–36. Durham, NC: Duke University Press.

Bell, Charles. 1928. *The People of Tibet*. Oxford: Clarendon Press.

———. 2000 [1924]. *Tibet Past and Present*. New Delhi: Motilal Banarsidass.

Bestor, Theodore. 2001. "Supply-Side Sushi: Commodity, Market, and the Global City." *American Anthropologist* 103(1): 76–95.

Bharat Online. N.d. "Nathula Pass." Electronic document, http://www.bharatonline.com/sikkim/travel/east-sikkim/nathula-pass.html, accessed March 20, 2008.

Bird, Jon, with Barry Curtis, Tim Putnam, George Robertson, and Lisa Tickner. 1993. *Mapping the Futures: Local Cultures, Global Change*. London: Routledge.

Bishop, Peter. 1989. *The Myth of Shangri-La: Tibet, Travel Writing, and the Western Creation of Sacred Landscape*. Berkeley: University of California Press.

Black, Jeremy. 2000. *Maps and Politics*. Chicago: University of Chicago Press.

Bloch, Maurice, and Jonathan P. Parry. 1990. "Introduction: Money and the Morality of Exchange." In *Money and the Morality of Exchange*, ed. Maurice Bloch and Jonathan P. Parry, 1–32. Cambridge: Cambridge University Press.

Boulnois, Luce. 2003. "Gold, Wool, and Musk: Trade in Lhasa in the Seventeenth Century." In *Lhasa in the Seventeenth Century: The Capital of the Dalai Lamas,* ed. Françoise Pommaret, 133–56. Leiden: Brill.

Bourdieu, Pierre. 1977. *Outline of a Theory of Practice*. Cambridge: Cambridge University Press.

Braudel, Fernand. 1977. *Afterthoughts on Material Civilization and Capitalism*. Baltimore: Johns Hopkins University Press.

———. 1992 [1982]. *The Wheels of Commerce: Civilization and Capitalism, 15th–18th Century*, vol. 3. Trans. Siân Reynolds. New York: Harper and Row.

Brenner, Neil. 1998. "Between Fixity and Motion: Accumulation, Territorial Organization, and the Historical Geography of Spatial Scales." *Environment and Planning D: Society and Space* 16(4): 459–81.

Buchli, Victor, ed. 2002. *The Material Culture Reader*. Oxford: Berg.

Cammack, Paul. 2010. "The Shape of Capitalism to Come." In *The Point Is to Change It: Geographies of Hope and Survival in an Age of Crisis*, ed. Noel Castree et al. Chichester: Wiley-Blackwell.

Cammann, Schuyler V. R. 1970. *Trade through the Himalayas: The Early British Attempts to Open Tibet*. Westport, CT: Greenwood.

Castells, Manuel. 1996. *The Rise of the Network Society*. Oxford: Blackwell.

Chaudhari, Kalyan. 2003. "Routes of Promise." *Frontline*, July 5–18: 20(14).

Cheater, Angela. 1998. "Transcending the State? Gender and Borderline Constructions of Citizenship in Zimbabwe." In *Border Identities: Nation and State at International Frontiers*, ed. Thomas W. Wilson and Hastings Donnan, 191–214. Cambridge: Cambridge University Press.

China Daily. 2006. "Silk Road Revived as Border Pass Reopens After 44 Years." Electronic document, http://english.peopledaily.com.cn, accessed July 23, 2006.

Christian, David. 2000. "Silk Roads or Steppe Roads? The Silk Roads in World History." *Journal of World History* 11(1): 1–26.

Clifford, James. 1997. *Routes: Travel and Translation in the Late Twentieth Century*. Cambridge, MA: Harvard University Press.

Cohen, Abner. 1969. *Custom and Politics in Urban Africa: A Study of Hausa Migrants in Yoruba Towns*. Berkeley: University of California Press.

Cohen, Erik. 2000. *The Commercialized Crafts of Thailand: Hill Tribes and Lowland Villages*. London: Curzon Press.

Cook, Ian, et al. 2004. "Follow the Thing: Papaya." *Antipode* 36(4): 642–64.

Cook, Ian, J. Evans, H. Griffiths, R. Morris, and S. Wrathmell. 2007. "'It's More Than Just What It Is': Defetishising Commodities, Expanding Fields, Mobilising Change . . ." *Geoforum* 38: 1113–26.

Cowen, Deborah. 2010. "A Geography of Logistics: Market Authority and the Security of Supply Chains." *Annals of the Association of American Geographers* 100(3): 600–620.

Curtin, Philip. 1984. *Cross-Cultural Trade in World History*. Cambridge: Cambridge University Press.

Das, Veena, and Deborah Poole. 2004. *Anthropology in the Margins of the State*. Santa Fe: School of American Research Press.

de Certeau, Michel. 2002. *The Practice of Everyday Life*. Berkeley: University of California Press.

Dhondrup, Lhakpa. 2000. *Drel pa'i Mi tshe (The Life of a Muleteer)*. Beijing: Beijing Minorities Publishing House.

Diehl, Keila. 2002. *Echoes from Dharamsala: Music in the Life of a Tibetan Refugee Community*. Berkeley: University of California Press.

Dixit, Kanak. 2002. "Chicken's Neck." *Himal Southasian* 15(8): 60.

Dodin, Thierry, and Heinz Räther, eds. 2001. *Imagining Tibet: Perceptions, Projections, and Fantasies*. Somerville, MA: Wisdom Publications.

Douglas, Mary. 2002. *Purity and Danger*. London: Routledge.

Douglas, Mary, and Baron C. Isherwood. 1996. *The World of Goods: Towards an Anthropology of Consumption*. New York: Routledge.

Economic and Political Weekly. 2005. "India and China Move Forward." *Economic and Political Weekly* 40(16): 1567.

Edney, Matthew. 1997. *Mapping an Empire: The Geographical Construction of British India, 1765–1843*. Chicago: University of Chicago Press.

Fader, H. Louis. 2002. *Called from Obscurity: The Life and Times of Gergan Tharchin*, vol. 1. Kalimpong: Tibet Mirror Press.

———. 2004. *Called from Obscurity: The Life and Times of Gergan Tharchin*, vol. 2. Kalimpong: Tibet Mirror Press.

Ferme, Marianne C. 2004. "Deterritorialized Citizenship and the Resonances of the Sierra Leonean State." In *Anthropology in the Margins of the State*, ed. Veena Das and Deborah Poole, 81–115. Santa Fe: School of American Research Press.

Fewkes, Jacqueline. 2005. "The Legacy of Trade: Social Networks in Ladakh, India." PhD dissertation, Department of Anthropology, University of Pennsylvania.

Fischer, Andrew. 2005. *State Growth and Social Exclusion in Tibet*. Copenhagen: Nordic Institute of Asian Studies Press.

Fisher, James F. 1997 [1986]. *Trans-Himalayan Traders: Economy, Society, and Culture in Northwest Nepal*. New Delhi: Motilal Banarsidass Publishers.

Foucault, Michel. 1979. "Governmentality." *Ideology and Consciousness* 6: 5–21.

Franquesa, Jaume. 2011. "'We've Lost Our Bearings': Place, Tourism, and the Limits of the 'Mobility Turn.'" *Antipode* 43(4): 1012–33.

Fuller, Thomas. 2008. "In Isolated Hills of Asia, New Roads to Speed Trade." *New York Times*, March 31.

Gereffi, Gary, and Miguel Korzeniewicz, eds. 1994. *Commodity Chains and Global Capitalism*. Westport, CT: Praeger.

Goldstein, Melvyn C., and Cynthia M. Beall. 1989. "The Impact of China's Reform Policy on Nomadic Pastoralists in Western Tibet." *Asian Survey* 29(6): 619–41.

Gordon, Avery. 1997. *Ghostly Matters: Haunting and the Sociological Imagination*. Minneapolis: University of Minnesota Press.

Graburn, Nelson H. H., ed. 1976. *Ethnic and Tourist Arts: Cultural Expressions from the Fourth World*. Berkeley: University of California Press.

Graeber, David. 2011. *Debt: The First Five Thousand Years*. New York: Melville House Books.

Grunfeld, A. Tom. 1996. *The Making of Modern Tibet*. Rev. edition. Armonk: M. E. Sharpe.

Gupta, Akhil, and James Ferguson, eds. 1997. *Culture, Power, Place: Explorations in Critical Anthropology*. Durham, NC: Duke University Press.

Handler, Richard. 1986. "Authenticity." *Anthropology Today* 2(1): 6–9.

Hansen, Karen Tranberg. 2000. *Salaula: The World of Secondhand Clothing and Zambia*. Chicago: University of Chicago Press.

Harris, Clare. 1999. *In the Image of Tibet: Tibetan Painting after 1959*. London: Reaktion Books.

Harris, Tina. 2007. "Towards a Geographical Anthropology of Trade in the Himalayas." In *Toronto Studies in Central and Inner Asia: Traders and Trade Routes of Central and Inner Asia, Then and Now*, ed. Michael Gervers, Uradyn Bulag, and Gillian Long, 189–206. Toronto: Central and Inner Asia Seminar.

———. 2008. "Silk Roads and Wool Routes: Contemporary Geographies of Trade between Lhasa and Kalimpong." *India Review* 7(3): 200–222.

———. 2012. "From Loom to Machine: Tibetan Aprons and the Configuration of Space." *Environment and Planning D: Society and Space* 30(5): 877–95.

Hart, Gillian. 1998. "Multiple Trajectories: A Critique of Industrial Restructuring and the New Institutionalism." *Antipode* 30(4): 333–56.

Harvey, David. 1990. *The Condition of Postmodernity*. Oxford: Blackwell.

———. 1991. "Between Space and Time: Reflections on the Geographic Imagination." *Annals of the Association of American Geographers* 80(3): 418–34.

———. 1999. *Limits to Capital*. London: Verso.

———. 2001. *Spaces of Capital: Towards a Critical Geography*. New York: Routledge.

———. 2003. *The New Imperialism*. Oxford: Oxford University Press.

Haugerud, Angelique, M. Priscilla Stone, and Peter D. Little, eds. 2000. *Commodities and Globalization: Anthropological Perspectives*. Lanham, MD: Rowman and Littlefield.

Hendrickson, Carol. 1996. "Selling Guatemala: Maya Export Products in US Mail-Order Catalogues." In *Cross-Cultural Consumption: Global Markets, Local Realties*, ed. David Howes, 106–21. New York: Routledge.

Henrion-Dourcy, Isabelle. 2005. "Women in the Performing Arts: Portraits of Six Contemporary Singers." In *Women in Tibet: Past and Present*, ed. Janet Gyatso and Hanna Havnevik, 195–258. New York: Columbia University Press.

Hepburn, Sharon. 2000. "The Cloth of Barbaric Pagans: Tourism, Identity, and Modernity in Nepal." *Fashion Theory* 4(3): 275–300.

Hilker, Deb Shova Kansakar. 2005. *Syamukapu: The Lhasa Newars of Kalimpong and Kathmandu*. Kathmandu: Vajra Publications.

Hridaya, Chittadhar. 2002. *Mimmanahpau: Letter from a Lhasa Merchant to His Wife*. Trans. Kesar Lall. New Delhi: Robin Books.

Hu, Xiaojiang. 2004. "The Little Shops of Lhasa, Tibet: Migrant Businesses and the Formation of Markets in a Transitional Economy." PhD dissertation, Department of Sociology, Harvard University.

Humphrey, Caroline. 1999. "Traders, 'Disorder,' and Citizenship Regimes in Provincial Russia." In *Uncertain Transition: Ethnographies of Change in the Postsocialist World*, ed. Michael Burawoy and Katherine Verdery, 19–52. Lanham, MD: Rowman and Littlefield.

Humphrey, Caroline, and Stephen Hugh-Jones, eds. 1992. *Barter, Exchange, and Value: An Anthropological Approach*. Cambridge: Cambridge University Press.

Hussain, Wasbir. 2006. "Silk Road Pass Reopened as China and India Build Closer Ties." *The Independent*, July 7.

Inda, Jonathan Xavier, and Renato Rosaldo, eds. 2001. "Introduction: A World in Motion." In *The Anthropology of Globalization: A Reader*, 1–35. Oxford: Blackwell.

India E-News. 2006. "History Made as Silk Road Is Reopened for Border Trade." Electronic document, http://www.indiaenews.com, accessed July 8, 2006.

India Office Records. 1945. Annual Report of the British Trade Agency, Yatung, Tibet, for the year ending 31st March 1945. London: India Office Records, British Library. L/P&S/12/4166, Pol Ext Coll 36 File 3.

———. 1946. Extract from Lhasa Letter ending 21.4.1946. London: India Office Records, British Library. L/P&S/12/4220, Ext 3769/46.

International Campaign for Tibet. 2003. "Crossing the Line: China's Railway to Lhasa, Tibet." Washington, DC: International Campaign for Tibet.

Johnson, Corey, Reece Jones, Anssi Paasi, Louise Amoore, Alison Mountz, Mark Salter, and Chris Rumford. 2011. "Interventions on Rethinking 'the Border' in Border Studies." *Political Geography* 30: 61–69.

Katz, Cindi. 1996. "Towards Minor Theory." *Environment and Planning D: Society and Space* 14(4): 487–99.

———. 2001. "On the Grounds of Globalization: A Topography for Feminist Political Engagement." *Signs: Journal of Women in Culture and Society* 26(4): 1213–35.

Kearney, Michael. 1991. "Borders and Boundaries of State and Self at the End of Empire." *Journal of Historical Sociology* 4(1): 52–74.

Kearns, Gerry. 2006. "Naturalising Empire: Echoes of Mackinder for the Next American Century?" *Geopolitics* 11(1): 74–98.

Kitchin, Robert M. 1994. "Cognitive Maps: What Are They and Why Study Them?" *Journal of Environmental Psychology* 14(1): 1–19.

Klieger, P. Christiaan. 1992. "Shangri-La and the Politicization of Tourism in Tibet." *Annals of Tourism Research* 19(1): 122–25

Klimburg, Maximilian. 1982. "The Western Trans-Himalayan Crossroads." In *The Silk Route and the Diamond Path: Esoteric Buddhist Art on the Trans-Himalayan Trade Routes*, ed. Deborah Klimburg-Salter, 25–37. Los Angeles: UCLA Art Council.

Konstantinov, Yulian. 1996. "Patterns of Reinterpretation: Trader-Tourism in the Balkans (Bulgaria) as a Picaresque Metaphorical Enactment of Post-Totalitarianism." *American Ethnologist* 23(4): 762–82.

Kopytoff, Igor. 1986. "The Cultural Biography of Things: Commoditization as Process." In *The Social Life of Things: Commodities in Cultural Perspective*, ed. Arjun Appadurai, 64–94. Cambridge: Cambridge University Press.

Krishna, Sankaran. 1996. "Cartographic Anxiety: Mapping the Body Politic in India." In *Challenging Boundaries: Global Flows, Territorial Identities*, ed. Michael J. Shapiro and Wayward R. Alker, 193–215. Minneapolis: University of Minnesota Press.

Lall, Kesar. 2001. *The Newar Merchants in Lhasa*. Kathmandu: Ratna Pustak Bhandar.

Lattimore, Owen. 1962. *Studies in Frontier History: Collected Papers, 1928–1958*. Paris: Mouton.

Lefebvre, Henri. 1991 [1974]. *The Production of Space*. Oxford: Blackwell.

———. 2003 [1970]. *The Urban Revolution*. Minneapolis: University of Minnesota Press.

Lewis, Todd. 2003. "Buddhist Merchants in Kathmandu: The Asan Twah Market and Uray Social Organization." In *Contested Hierarchies: A Collaborative Ethnography of Caste among the Newars of the Kathmandu Valley, Nepal* ed. David Gellner and Declan Quigley, 38–79. Oxford: Oxford University Press.

Li, Dezhu. 2000. "Large-Scale Development of Western China and China's Nationality Problem." *Qiushi (Seeking Truth)*, June 15.

Lionnet, Françoise, and Shu-mei Shih, eds. 2005. *Minor Transnationalism*. Durham, NC: Duke University Press.

Löfgren, Orvar. 1999. "Crossing Borders: The Nationalization of Anxiety." *Ethnologia Scandinavica* 29: 5–27.

Ludden, David. 2003. "Presidential Address: Maps in the Mind and the Mobility of Asia." *Journal of Asian Studies* 62(4): 1057–78.

Ma, Lihua. 2003. *Old Lhasa: A Sacred City at Dusk*. Beijing: Foreign Languages Press.

MacDonald, David. 1999 [1930]. *Touring in Sikkim and Tibet*. New Delhi: Asian Educational Services.

———. 2005 [1938]. *Twenty Years in Tibet*. Varanasi: Pilgrims.

Madan, P. L. 2004. *Tibet: Saga of Indian Explorers (1864–1894)*. New Delhi: Manohar.

Makley, Charlene E. 2003. "Gendered Boundaries in Motion: Space and Identity on the Sino-Tibetan Frontier." *American Ethnologist* 30(4): 595–617.

Malinowski, Bronislaw. 1967 [1922]. *Argonauts of the Western Pacific: An Account of Native Enterprise and Adventure in the Archipelagoes of Melanesian New Guinea*. London: Routledge and Kegan Paul.

Marcus, George. 1995. "Ethnography in/of the World System: The Emergence of Multi-Sited Ethnography." *Annual Review of Anthropology* 24: 95–117.

Marx, Karl. 1990 [1867]. *Capital*, vol. 1. London: Penguin Classics.

———. 1992 [1893]. *Capital*, Vol. 2. London: Penguin.

Massey, Doreen. 1993. "Power-Geometry and a Progressive Sense of Place." In *Map-*

ping the Futures: Local Cultures, Global Change, ed. John Bird, Barry Curtis, Tim Putnam, and Lisa Tickner, 59–69. New York: Routledge.

———. 2002. "Globalisation as Geometries of Power." Electronic document, http://www.signsofthetimes.org.uk/massey%5Btextonly%5D.html, accessed June 21, 2008.

Massumi, Brian. 2002. *Parables for the Virtual: Movement, Affect, Sensation*. Durham, NC: Duke University Press.

Mauss, Marcel. 1954 [1925]. *The Gift: Forms and Functions of Exchange in Archaic Societies*. Glencoe: Free Press.

Maxwell, Neville. 1970. *India's China War*. New York: Pantheon.

McDuie-Ra, Duncan. 2011. "The Dilemmas of Pro-Development Actors: Viewing State-Ethnic Minority Relations and Intra-Ethnic Dynamics through Contentious Development Projects." *Asian Ethnicity* 12(1): 77–100.

McGranahan, Carole. 2001. "Arrested Histories: Between Empire and Exile in Twentieth Century Tibet." PhD dissertation, Department of Anthropology and History, University of Michigan.

———. 2002. "Sa sPang mda' gNam sPang mda': Murder, History, and Social Politics in 1920s Lhasa." In *Khams pa Local Histories: Visions of People, Place, and Authority*, ed. Lawrence Epstein, 103–26. Leiden: Brill.

McGuckin, Eric. 1997. "Tibetan Carpets: From Folk Art to Global Commodity." *Journal of Material Culture* 2(3): 291–310.

McKay, Alex. 1997. *Tibet and the British Raj: The Frontier Cadre, 1904–1947*. Richmond: Curzon.

McMurray, David A. 2003. "Recognition of State Authority as a Cost of Involvement in Moroccan Border Crime." In *Crime's Power: Anthropologists and the Ethnography of Crime*, ed. Philip Parnell and Stephanie Kane, 125–44. New York: Palgrave Macmillan.

Middleton, John. 2003. "Merchants: An Essay in Historical Ethnography." *Journal of the Royal Anthropological Institute* 9(3): 509–27.

Miller, Daniel. 1987. *Material Culture and Mass Consumption*. Oxford: Blackwell.

———. 1995. "Consumption and Commodities." *Annual Review of Anthropology* 24: 141–61.

Mintz, Sidney. 1985. *Sweetness and Power: The Place of Sugar in Modern History*. New York: Viking.

Mitchell, Don. 1996. *The Lie of the Land: Migrant Workers and the California Landscape*. Minneapolis: University of Minnesota Press.

Mitchell, Timothy. 1999. "Society, Economy, and the State Effect." In *State/Culture: State-Formation after the Cultural Turn*, ed. George Steinmetz, 76–97. Ithaca, NY: Cornell University Press.

Murakami, Daisuke. 2007. "Tourism Development and Propaganda in Contemporary Lhasa, Tibet Autonomous Region (TAR), China." In *Asian Tourism: Growth and Change*, ed. Janet Cochrane, 55–68. New York: Elsevier Science.

Myers, Fred. 2001. "Introduction." In *The Empire of Things: Regimes of Value and Material Culture*, ed. Fred Myers, 3–61. Santa Fe: School of American Research Press.

Newman, Robert P. 1992. *Owen Lattimore and the "Loss" of China*. Berkeley: University of California Press.

New York Times. 1951a. "Tibetan Wool Soars on American Demand." *New York Times*, January 22.

———. 1951b. "Tibetan Trade Cut by Rupee Scarcity." *New York Times*, December 25.

———. 1952. "Peiping to Buy Tibet's Wool." *New York Times*, May 23.

———. 1964. "Nepal Road Aims at Link to Tibet." *New York Times*, February 9.

Nordstrom, Carolyn. 2007. *Global Outlaws: Crime, Money, and Power in the Contemporary World*. Berkeley: University of California Press.

Ong, Aihwa. 1999. *Flexible Citizenship: The Cultural Logics of Transnationality*. Durham, NC: Duke University Press.

Pak-China Sust Port. N.d. "Pak-China Sust Port." Electronic document, http://sustdryport.com, accessed June 1, 2008.

Pickles, John. 2004. *A History of Spaces: Cartographic Reason, Mapping, and the Geo-Coded World*. London: Routledge.

Plattner, Stuart. 1996. *High Art Down Home: An Economic Ethnography of a Local Art Market*. Chicago: University of Chicago Press.

Polanyi, Karl. 2001 [1944]. *The Great Transformation*. Boston: Beacon.

Pratt, Mary Louise. 1992. *Imperial Eyes: Travel Writing and Transculturation*. New York: Routledge.

Rabinow, Paul, and George Marcus, with James D. Faubion and Tobias Rees. 2008. *Designs for an Anthropology of the Contemporary*. Durham, NC: Duke University Press.

Radhu, Abdul Wahid. 1997. *Islam in Tibet: The Illustrated Tibetan Caravans*. Louisville, KY: Fons Vitae.

Rafael, Vicente. 1997. "'Your Grief Is Our Gossip': Overseas Filipinos and Other Spectral Presences." *Public Culture* 9: 267–91.

Rajesh, Y. P. 2006. "Hostility Put Aside, India and China Open Silk Road." *Guardian Weekly*, July 14–20.

Ramachandran, Sudha. 2008. "Nepal to Get China Rail Link." Electronic document, http://www.atimes.com/atimes/South_Asia/JE15Df01.html, accessed July 23, 2008.

Ramesh, Randeep. 2006. "Book-Burning Threat over Town's Portrayal in Booker-Winning Novel." *Guardian*, November 6: 23.

Ratanapruck, Prista. 2008. "Manangi Trade Diasporas in South Asia and Southeast Asia." Paper presented at the Annual Meeting of the Association for Asian Studies, Atlanta, April 3–8.

Reeves, Madeline. 2008. "Materializing Borders." *Anthropology News*. 49(5): 12–13.

Reuters. 2008. "Why Is Remote Tibet of Strategic Significance?" Electronic document, http://www.reuters.com/article/latestCrisis/idUSSP23050, accessed March 25, 2008.

Rizvi, Janet. 1999. *Trans-Himalayan Caravans: Merchant Princes and Peasant Traders in Ladakh*. New Delhi: Oxford University Press.

Robin, Ravinder Singh. 2006. "Border Baba's Blessing for Nathu La Trade Cheers Amritsar Devotees." Electronic document, http://www.newkerala.com, accessed June 27, 2007.

Saha, Narayan Chandra. 2003. *The Marwari Community in Eastern India*. New Delhi: Decent Books.

Samuel, Geoffrey. 1994. "Tibet and the Southeast Asian Highlands: Rethinking the Intellectual Content of Tibetan Studies." In *Tibetan Studies: Proceedings of the Sixth Seminar of the International Association of Tibetan Studies*, ed. Per Kvaerne, 696–710. Oslo: Institute for Comparative Research in Human Culture.

Sassen, Saskia. 2002. *Global Networks, Linked Cities*. London: Routledge.

Sautman, Barry, and June Dreyer, eds. 2006. *Contemporary Tibet: Politics, Development and Society in a Disputed Region*. Armonk, NY: M. E. Sharpe.

Schafer, Edward H. 1985. *The Golden Peaches of Samarkand: A Study of T'ang Exotics*. Berkeley: University of California Press.

Schivelbusch, Wolfgang. 1986. *The Railway Journey: The Industrialization of Time and Space in the Nineteenth Century*. Berkeley: University of California Press.

Schwartz, Ronald D. 1995. *Circle of Protest: Political Ritual in the Tibetan Uprising, 1987–1992*. New York: Columbia University Press.

Scott, James. 2009. *The Art of Not Being Governed: An Anarchist History of Upland Southeast Asia*. New Haven, CT: Yale University Press.

Secor, Anna. 2007. "Between Longing and Despair: State, Space, and Subjectivity in Turkey." *Environment and Planning D: Society and Space* 24: 33–52.

Shakabpa, Tsepon W. D. 1967. *Tibet: A Political History*. New Haven, CT: Yale University Press.

Shakya, Tsering. 1990. "1948 Tibetan Trade Mission to United Kingdom." *Tibet Journal* 15(4): 97–114.

———. 1993. "Whither the Tsampa Eaters?" *Himal Southasian* 6(5): 8–11.

———. 1999. *The Dragon in the Land of Snows: A History of Modern Tibet since 1947*. London: Pimlico.

———. 2008. "Interview: Tibetan Questions." *New Left Review* 51: 5–26.

Shields, Rob. 1991. *Places on the Margin: Alternative Geographies of Modernity*. London: Routledge.

Shneiderman, Sara. 2005. "Swapping Identities: Borderland Exchanges along the Nepal-TAR Border." *Himal Southasian* 18(3): 32–33.

———. 2010. "Are the Central Himalayas in Zomia? Some Scholarly and Political Considerations across Time and Space." *Journal of Global History* 5(2): 289–312.

Simmel, Georg. 1990 [1907]. *The Philosophy of Money*, 2nd ed. Trans. Tom Bottomore and David Frisby. New York: Routledge.

Smith, Neil. 2003. Foreword to *The Urban Revolution*, by Henri Lefebvre, 2003 [1970], vii–xxiii. Minneapolis: University of Minnesota Press.

———. 2008 [1984]. *Uneven Development: Nature, Capital, and the Production of Space.* Athens: University of Georgia Press.

Snellgrove, David, and Hugh Richardson. 1968. *A Cultural History of Tibet.* Boston: Shambhala.

Spooner, Brian. 1986. "Weavers and Dealers: Authenticity and Oriental Carpets." In *The Social Life of Things: Commodities in Cultural Perspective*, ed. Arjun Appadurai, 195–235. Cambridge: Cambridge University Press.

Steiner, Christopher B. 1994. *African Art in Transit.* Cambridge: Cambridge University Press.

Stephen, Lynn. 2007. *Transborder Lives: Indigenous Oaxacans in Mexico, California, and Oregon.* Durham, NC: Duke University Press.

Subba, Tanka. 1990. *Flight and Adaptation: Tibetan Refugees in the Darjeeling-Sikkim Himalaya.* New Delhi: Library of Tibetan Works and Archives.

Taussig, Michael. 1997. *The Magic of the State.* New York: Routledge.

———. 2004. *My Cocaine Museum.* Chicago: University of Chicago Press.

Thomas, Nicholas. 1991. *Entangled Objects: Exchange, Material Culture and Colonialism in the Pacific.* Cambridge: Harvard University Press.

Thomas, Samuel. 2002. "Saint-Sentinel: Harbhajan Singh of Upper Sikkim." *Himal Southasian* 15(3): 50–53.

Toyota, Mika. 2000. "Akha Border Trade." In *Where China Meets Southeast Asia: Social and Cultural Change in the Border Regions*, ed. Grant Evans, Christopher Hutton, and Kuah Khun, 204–21. Singapore: Institute of Southeast Asian Studies.

Tsing, Anna. 2005. *Friction: An Ethnography of Global Connection.* Princeton, NJ: Princeton University Press.

Tuladhar, Kamal. 2004. *Caravan to Lhasa: Newar Merchants of Kathmandu in Traditional Tibet.* Kathmandu: Tuladhar Family.

Turin, Mark, and Sara Shneiderman. 2003. "Yams in Boulderland: Nepalis on the Plateau." *Nepali Times*, September: 8–9.

Ubyssey. 1958. "Shortage of Yak Tails Worries Santa." *Ubyssey*, November 4: 5.

van Schendel, Willem. 2002. "Geographies of Knowing, Geographies of Ignorance: Jumping Scale in Southeast Asia." *Environment and Planning D: Society and Space* 20: 647–68.

———. 2005. "Spaces of Engagement: How Borderlands, Illegal Flows, and Territorial States Interlock." In *Illicit Flows and Criminal Things: States, Borders, and the Other Side of Globalization*, ed. Itty Abraham and Willem van Schendel, 38–68. Bloomington: Indiana University Press.

van Spengen, Wim. 2000. *Tibetan Border Worlds: A Geo-Historical Analysis of Trade and Traders.* London: Kegan Paul.

Vann, Elizabeth F. 2006. "The Limits of Authenticity in Vietnamese Consumer Markets." *American Anthropologist* 108(2): 286–95.

Verdery, Katherine. 1996. "The Elasticity of Land: Problems of Property Restitution in Transylvania." In *What Was Socialism and What Comes Next?* ed. Katherine Verdery, 133–67. Princeton, NJ: Princeton University Press.

von Fürer-Haimendorf, Christoph. 1975. *Himalayan Traders: Life in Highland Nepal.* London: John Murray.

Walker, Andrew. 1999. *The Legend of the Golden Boat: Regulation, Trade, and Traders in the Borderlands of Laos, Thailand, China, and Burma.* Honolulu: University of Hawaii Press.

Wallerstein, Immanuel. 2000. *The Essential Wallerstein.* New York: New Press.

White, Bob. 2000. "Soukouss or Sell-Out? Congolese Popular Music as Cultural Commodity." In *Commodities and Globalization: Anthropological Perspectives*, ed. Angelique Haugerud, M. Priscilla Stone, and Peter D. Little, 33–58. Lanham, MD: Rowman and Littlefield.

Williams, Raymond. 1961. *The Long Revolution.* London: Chatto and Windus.

———. 1985 [1973]. *The Country and the City.* London: Hogarth.

Wilson, Thomas W., and Hastings Donnan, eds. 1998. *Border Identities: Nation and State at International Frontiers.* Cambridge: Cambridge University Press.

Winichakul, Thongchai. 1994. *Siam Mapped: Maps and the Formation of the Geo-Body in Siam.* Honolulu: University of Hawai'i Press.

Wolf, Eric. 1982. *Europe and the People without History.* Berkeley: University of California Press.

Xinhua. 2008. "Tibet Plans Huge Industrial Investment." Electronic document, http://eng.tibet.cn/news/today/200809/t20080907_424824.htm, accessed January 24, 2009.

Yeh, Emily. 2003. "Taming the Tibetan Landscape: Chinese Development and the Transformation of Agriculture." PhD dissertation, Department of Energy and Resources, University of California, Berkeley.

———. 2006. "'An Open Lhasa Welcomes You': Disciplining the Researcher in Tibet." In *Doing Fieldwork in China*, ed. Maria Heimer and Stig Thøgersen, 96–109. Honolulu: University of Hawai'i Press.

———. 2007a. "Transnational/Translocal Collaboration and Environmental Subject Formation: Sacred Lands and the Green Tibetan." Paper presented at the Annual Meeting of the American Anthropological Association, Washington, DC, November 29.

———. 2007b. "Tropes of Indolence and the Cultural Politics of Development in Lhasa, Tibet." *Annals of the Association of American Geographers* 97(3): 593–612.

———. 2009. "From Wasteland to Wetland? Nature and Nation in China's Tibet." *Environmental History* 14(1): 32–66.

Yeh, Emily, and Mark Henderson. 2008. "Interpreting Urbanization in Tibet: Administrative Scales and Discourses of Modernization" *Journal of the International Association of Tibetan Studies*, No. 4, http://www.thlib.org/collections/texts/jiats/#!jiats=/04/yeh/b5/

Younghusband, Francis. 1996 [1910]. *India and Tibet: A History of the Relations which Have Subsisted Between the Two Countries from the Time of Warren Hastings to 1910, With A Particular Account of the Mission to Lhasa of 1904*. Ed. Robert Bamford. Golden Gale Electronic Books.

Ziegler, Catherine. 2004. "Favored Flowers: Culture and Markets in a Global Commodity Chain." PhD dissertation, Department of Anthropology, New School University.

INDEX

Page numbers in italics indicate figures and maps.

"big" or "small" and, 118–19; maps of, 84–86; social history of, 7–13, *8, 9, 10, 11*, 18. *See also* Tibet; *and specific cities, regions, and states*
Lhokha prefecture, 57–58, *58*, 60–61
Lingtam, 41–42

MacDonald, David, 34–35
macro-level research, 4, 31, 104, 140, 147–48
Maddin, Guy, 3
Mahendra (king), 48
maps: alternative, 2–3; British trade in the Himalayan region and, 8–9, 32–35; place or place-based narratives and, 19–20; spatial representations of regions and, 84–86
Marcus, George, 24, 26
Marwari traders, 12, 14–15, 110, *111*, 112
Marx, Karl, 6–7, 52, 54–56, 81, 88, 114
McGuckin, Eric, 66
McMurray, David, 104–5
Mela festival ground, *85*, 84–85
men: globalization in context of fashions for, 156; mule caravans and, 10, 27–28, 35–36, *36*, 39, 41–46, *43*; production of goods and, 82, 162. *See also* gender differences; identity of traders as "big" or "small"; traders; transportation developments; women
methodology and language, xv, 1–2, 24–26, 153–56, 157n1, 160n2
mi ba ru ("middleman"), 42–43. *See also* traders exchanges
micro-level research, 4, 140
"middleman" (*mi ba ru*), 42–43. *See also* traders exchanges
Middleton, John, 16
migrant workers, in *pang gdan* production, 61–62
Mintz, Sidney, 53
Mitchell, Don, 157n2

Mitchell, Timothy, 96
mobility and fixity interplay: summary of, 4–5, 23, 150–52; diversions and, 150–52; economic geographies and, 124–27; everyday life experiences and, 141–42; gender differences and, 10; global economic processes and, 147–48; globalization in context of, 87–89, 150–52; spatial representations of regions and, 87–89; spatio-temporal tensions in context of, 92–93, 96, 160n4; tourists and, 146, 148; trade in historical context and, 6, 146–50. *See also* fixing practices
modern consumers, 62–63. *See also* traders exchanges
mule caravans, 10, 27–28, 35–36, *36*, 39, 41–46, *43*. *See also* traders; transportation developments
multi-sited (multi-sitedness), 24, 26. *See also* methodology and language

Nagatse, 41–42
Nanga, 41–42
Nathu-la: blind fields concept and, 145; Border Baba haunting and, 47, 49, 50–51; border crossing described at, 100; borders as zones of contradictions and, 107–9; environmental discourse and, 148; geopolitical history of trade and, 13, 20–22, 30, 47; identity of traders as "big" or "small" and, 119–20; "opening" or "freeing" border trade and, 107; permits for trading and, 100, 107, 111; reconnections across borders and, 109–12, *111*; reopening of, *85*, 89–90, *91*, 110, 160n3; social history of Lhasa to Kalimpong route and, 9, *11*, 13; state-level policies and, 107
nation-state: borders and, 50, 77; border studies and, 102, 105; China as, 77, 89, 93, 123, 145, 149–50; globalization

GEOGRAPHIES OF JUSTICE AND SOCIAL TRANSFORMATION